RENEWALS 458-4574
DATE DUE

WITHDRAWN
UTSA LIBRARIES

Growing Global

GROWING GLOBAL

A Corporate Vision Masterclass

Stan Shih

Edited by Fu-Yuan Xiao

Translated by Minn Song

John Wiley & Sons (Asia) Pte Ltd
Singapore   New York   Chichester
Brisbane   Toronto   Weinheim

Copyright © 2002 John Wiley & Sons (Asia) Pte Ltd
Published by John Wiley & Sons (Asia) Pte Ltd
2 Clementi Loop, #02-01, Singapore 129809

All rights reserved.

No part of this publication may be reproduced, stored in a retrieval system, or transmitted in any form or by any means, electronic, mechanical, photocopying, recording, scanning, or otherwise, except as expressly permitted by law, without either the prior written permission of the Publisher, or authorization through payment of the appropriate photocopy fee to the Copyright Clearance Center. Requests for permission should be addressed to the Publisher, John Wiley & Sons (Asia) Pte Ltd, 2 Clementi Loop, #02-01, Singapore 129809, tel: 65-463-2400, fax: 65-463-4605, E-mail: enquiry@wiley.com.sg.

This publication is designed to provide accurate and authoritative information in regard to the subject matter covered. It is sold with the understanding that the Publisher is not engaged in rendering professional services. If professional advice or other expert assistance is required, the services of a competent professional person should be sought.

Other Wiley Editorial Offices

John Wiley & Sons, Inc., 605 Third Avenue, New York, NY 10158-0012, USA
John Wiley & Sons Ltd, Baffins Lane, Chichester, West Sussex PO19 1UD, England
John Wiley & Sons (Canada) Ltd, 22 Worcester Road, Rexdale, Ontario M9W 1L1, Canada
John Wiley & Sons Australia Ltd, 33 Park Road (PO Box 1226), Milton, Queensland 4046, Australia
Wiley-VCH, Pappelallee 3, 69469 Weinheim, Germany

Library of Congress Cataloging-in-Publication Data:

Shi, Zhenrong, 1944–
    Growing global: A corporate vision masterclass / Stan Shih; edited by Fu-Yuan Xiao; translated by Minn Song.
       p. cm.
    Translated and collected from the author's lectures at the Chiaotung University.
    Includes index.
    ISBN 0-471-47927-6 (cloth : alk. paper)
      1. International business enterprises—Management. 2. Knowledge management—Taiwan. 3. Technological innovations—Taiwan—Management. 4. Industrial management—Taiwan. 5. Hong qi ji tuan (Taiwan) 6. Computer industry—Taiwan—Management—Case studies. I. Xiao, Fu-Yuan. II. Title.

HD62.4 .2523 2001
658'.049–dc21

2001026892

Typeset in 11/14 point, Minion by Linographic Services Pte Ltd
Printed in Singapore by Craft Print International Pte Ltd
10 9 8 7 6 5 4 3 2 1

Contents

Foreword — vii

Preface — xi

Part 1 The Knowledge Century

1. Paradigm shift — 3
2. Using small size to win big—The route to success in the Internet economy — 21
3. Internet organizations—Organizations tuned for the knowledge century — 37
4. A new vision—High-tech with a human touch — 59

Part 2 Market Share for Created Value

5. A new basis for competitiveness—Innovalue — 69
6. Eliminating the obstacles to innovation — 89

Part 3 Globalization Alternatives

7. An Asian approach to globalization — 113
8. Acer's breakthrough in the international arena — 133
9. From OEM to OBM — 153

Part 4 e-Leadership

10. Soft infrastructure—Vision and corporate culture — 177
11. Virtual dream teams — 195
12. New people or new thinking? — 211

References — 231

Index — 233

FOREWORD

Stan Shih is one of the Taiwanese businesspeople I most admire, and is also one of the businesspeople most respected by others in the business world and by aspirants still in school, both in Taiwan and abroad. The company he founded, Acer, is one of the few Taiwanese companies that has established its own brand and successfully entered the international market, all the while constantly innovating. It has also produced many high-level executives for high-tech companies.

Mr Shih has expressed many times in public that his greatest ambition as a student was to become a university professor. He accepted our invitation to conduct his own class at his alma mater, Chiaotung University, and share the invaluable experiences and insights gained from leading the internationalization of the Acer Group. Apart from giving Mr Shih a chance to return to his alma mater and fulfill a wish, the course was a great stroke of luck for the university's EMBA students who chose the course. I am delighted that through the publication of this book, readers around the world will be able to benefit from the ideas of this great businessperson and humanitarian.

The International Business Practices and Strategy course that Mr Shih subsequently gave at Chiaotung University, and on which the present book is based, encompassed four related broad topics: The Knowledge Century, Created Value Market Share, An Alternative Approach to Globalization, and e-Leadership. Mr Shih's course lectures included Acer's practices and strategies—past, present, and future. Since it was established, Acer has used small size to its advantage, and its real-world experience in internationalization serves as both a model and an inspiration for small and medium-size businesses.

After developing into a large corporation, Acer undertook a series of re-engineering initiatives in response to changes both within and outside the organization. Mr Shih sees the high speed of change in the high-tech industry as a fundamental factor in decision-making, and during the course he described in detail how Acer stayed on top of larger trends toward discontinuous technology development and industry super-disintegration. Finally, he described the sort of understanding that companies should have as they try to take advantage of new opportunities afforded by the knowledge era and Internet society, as well as suggestions for strategic responses from the country and business.

When the course first started, Mr Shih highlighted developmental trends in the international market and the industry paradigm shift by pointing out such phenomena as borderless markets, super-disintegration, and the digital economy, and their influence on business practices. The smiling curve that is familiar to people in the business world and the general public in parts of Asia graphically shows the relationship between the value chain and value creation that the evolution of the global PC industry has fostered. Mr Shih is adept at using an easy-to-understand approach to communicate complex ideas, and the smiling curve is the best example.

Regarding the knowledge economy and business opportunities on the Web, Mr Shih gave his personal views on knowledge management, new vision, development strategy, innovative ideas, and managerial concepts for organizational design. Of these subjects, the conceptual framework for an internet Organization Protocol (iOP) excited enthusiastic discussion among students and is extremely worthy of attention in academic and business circles.

Furthermore, on heeding the lessons of the smiling curve, the Acer Group's growth strategy has focused on customer-centric services and key components business, giving rise to another important perspective: the importance of innovation and the environment needed for the innovation process. That competitiveness is a function of the ratio between the so-called innovalue (added value created through innovation) and cost points up the importance of working toward strengthening capabilities in the area of innovalue. Innovation includes development in business models, technology, products, marketing, services, and supply chains, and Mr Shih detailed what the internal and external factors required by innovation mechanisms are and discussed them one by one.

The topic "An Alternative Approach to Globalization" explored the similarities and differences among American, European, and Japanese companies' approaches to internationalization, and also made evident the problems faced by Taiwanese business. Mr Shih, after considering the problems of Taiwanese companies, such as poor international image and insufficient personnel and financial resources, developed Acer's unique strategic approach to internationalization—using small size to win big. In the course, he compared the differences between OEM (original equipment manufacturing) and OBM (own-brand manufacturing) development strategies, explaining the important concepts and practices behind Acer's ability to institute total brand management in the face of so many products and markets—thus providing a valuable reference for business people and scholars.

The fourth section of the course, on e-leadership, included material on what vision and culture are, why they are important, and how to develop a corporate vision and firmly establish a corporate culture. In addition, this portion took examples from Acer's experience to explain the development strategies needed to realize an organizational vision, as well as re-engineering strategies and processes. Mr Shih's willingness to leave out nothing, teaching everything step by step, won the admiration of students.

Professor Buo-yong Ju
Department of Science Management
Chiaotung University
July 2001

PREFACE

This book represents my personal reflections as a CEO in building and managing one of the most recognized Asian global firms: the Acer Group. Founded in 1976 with US$25,000, the company has grown into one of the world's top ten PC companies and a diversified information technology conglomerate, designing and manufacturing a wide range of products such as PC peripherals, LCD flat panel displays, and communication products. The total revenue of the Acer Group in 2000 was US$9.4 billion, and it employs approximately 35,000 persons worldwide. In retrospect, our growth clearly represents a journey of continuous experimentation and learning as we have increased our global presence to include more than forty countries.

Though there are many business and management books written by outstanding CEOs, consultants, and scholars in the U.S. and Europe, there is a lack of business literature written from an Asian perspective. As an Asian CEO building a global firm with a very small home market and a relatively limited pool of global talent and financial resources, I have discovered that many strategies and experiences of the most admired U.S. or European firms cannot be directly applied without fully considering Asia's unique business conditions and resource availability. Had I followed the same strategies and growth patterns of our U.S. or European counterparts, our chances of winning in the global business arena would have been very limited. We must innovate and develop creative strategies that can leverage our strengths and compensate for our weaknesses.

As an Asian firm, Acer had to face four major challenges as we globalized. First, we needed to overcome the stereotype of "MIT" (Made in Taiwan) as we marketed and sold our products to millions of consumers who still perceived Taiwan as a place for manufacturing low-cost, low-quality products. This challenge is especially difficult for companies such as Acer that choose to develop their own brand names and market their products worldwide.

Second, we had very limited financial resources to fund global operations. The capital market in Taiwan is relatively small compared with those of the U.S. and Europe. As a result, even though we have been the number one brand of PC in Taiwan and South-east Asia for three consecutive years—1998, 1999, and 2000 (rated by *Reader's Digest*)—our ability to raise funds in Taiwan and South-east Asia is still relatively limited when compared with U.S. or European firms in their capital markets. As a result, we needed to be very careful in how we allocated our scarce resources when deciding which markets to penetrate and which products to develop.

Third, we had access to a relatively limited pool of global talent. Unlike the U.S. and Europe, where companies have a longer history of global operations, many firms in Asia have just started their globalization process. As a result, they have limited access to experienced executives who have developed a strong global mindset and management skills.

Fourth, many markets in Asia are highly fragmented and cannot enjoy the same economies of scale as in the U.S. or Pan-European markets. The PC industry in Taiwan, for example, accounts for less than 1% of the world market. Even though we have a very strong position in the Taiwanese market, much work is required to adapt our products, service infrastructures, and management practices for other major markets in order to build economies of scale. The scenario is very different for our leading competitors in the U.S., who start with a huge home market (the U.S. accounts for almost 40% of the worldwide PC market) and can enjoy economies of scale in their operations without going overseas. To them, going global requires incremental modifications rather than major changes.

While technologies such as those in PCs are global, marketing and management must be local. To respond to our constraints and challenges, at Acer we adopted three major globalization strategies. First, we used the "Global Brand, Local Touch" strategy to develop a strong partnership with leading local channel partners so that we could increase our access to funds, talent, and marketing capabilities in different strategic markets. Second, we used the "From Peripheral to Core" strategy, starting with secondary markets (for example, developing economies) where we could afford enough resources to engage in direct competition with other vendors, or offering peripheral products in core markets (such as the U.S. and Japan) where our competitors had already commanded a very strong presence. Third, our "Innovation in Technology and Management" strategy entailed the development of innovative products (such as Aspire) or management practices (such as employee stock ownership, fast-food business models, a client–server corporate structure, and internet organizations) that helped promote an innovative and progressive image (to combat the "MIT" stereotype). As a result of implementing these strategies, we were very pleased to learn that the *World Executive Digest* (March 1994) recognized the Acer approach of globalization as the "Fourth Way of Globalization," that is, different from the U.S., European, and Japanese approaches.

You may find my perspective on globalization somewhat different from those found in other mainstream management books written by U.S. or European authors. The value of this book comes from the ideas developed based on practical business and management conditions faced by small Asian firms that strive to become global players in the information technology industry. These ideas and lessons are extremely precious as they have been gained from paying hundreds of millions of dollars of "tuition" (that is, business mistakes). For example, it was only after many wrong turns had been taken that it finally became very clear to me that the future growth potential of Acer lay in strategic markets where we had a strong *local touch* (for example, the Chinese and Southeast Asian markets, where we enjoy language and cultural proximity, or other strategic markets like Europe, where we have a very strong local management team in place). It also took us many years to realize that our strategy to promote core products such as PC systems in the core markets (for example, in the U.S. or Japan) is not sustainable. Finally, the future of Acer is based on innovation in technologies and management that is distinct from the "me-too" style. In short, these ideas are not written as abstract concepts but as real business principles that guide our everyday business decisions and activities.

Growing Global was written for both Asian and non-Asian executives. By openly sharing the successes and failures of Acer, I hope this book can help reduce the learning curve for other Asian firms. A hopeful indication that this may prove to be the case is that the experience of Acer has provided a useful reference for Mr. Liu Chuanzhi, CEO of the Legend Group, which is currently the leading PC company in China, to stay focused in establishing a strong presence in its huge home market before going global. This book is also written for non-Asian executives who want to learn more about the business context and challenges that their Asian business partners are facing. By better appreciating why and how their Asian business partners operate, I hope non-Asian executives can develop more successful partnerships and alliances with their Asian counterparts.

This book does not offer any immediate solutions to the problems of globalization. It serves to stimulate thinking and raise further questions.

My sincere hope is that you will find this book valuable as we all learn to succeed in a more integrated global economy.

Stan Shih
July 2001

PART 1

THE KNOWLEDGE CENTURY

CHAPTER 1

Paradigm Shift

COMPUTER ARCHITECTURES have evolved from mainframes to the Internet, while the structure of the industry itself has changed from one stressing vertical integration to a new mode called "super-disintegration" based on a fine division of labor among highly interdependent companies.[1] With the dominance of Internet protocols, global standards have become open, and the world has come to depend heavily on them.

Based on recent trends, the development of the current global industry shows six major trends:

- increasingly larger and freer markets
- a borderless market economy
- development of super-disintegration
- a shift from product-centric to customer-centric
- a shift in the sources of value creation
- the digital revolution in the information era

INCREASINGLY LARGER AND FREER MARKETS

The various efforts made by businesses are all aimed at creating additional markets; and as markets grow more open and free, the benefits

---

1 The term "disintegration" was originally coined to describe the evolution in the computer industry away from large vertically integrated corporations toward a large number of highly specialized companies that each played a particular role, but with all roles tightly interdependent; "super-disintegration" refers to further refinement of this process, with even greater specialization.

of these efforts will be enhanced. Prior to the 1990s, many countries regarded protectionism as beneficial to themselves, believing that their economies could be self-sufficient, while in fact this thinking is misguided. Using Taiwan as an example, at one time the local market for home appliances was protected, making it very difficult for imported products to establish themselves, with the eventual result that the Taiwanese home appliance industry never developed international competitiveness.

As for information technology products in Taiwan, duties on imported finished goods are currently 5%, while the tax on imported components is 20%. With the help of this anti-protectionist policy, Taiwan's information technology industry has developed international competitiveness. The facts show that prolonged protectionism has no successful precedents, though for industries in the beginning stages of development, there is nothing wrong with the early support and guidance provided by protectionist policies. The situation can be likened to rearing offspring: nurturing provided during childhood is desirable, but permanent coddling and protectiveness prevents development into mature, robust adulthood.

Although it may be operating under a system of free competition, if a business can win, the result may be a high level of real rewards. However, in the final analysis, the nature of the victory must be considered. If the victory won is short term and can't be sustained over the long term, change is the key. If a business is only able to win in the short term, will it be able to win if the rules of the game change—if, for example, because of changes in the market, things that were highly valued in the past lose their value? In such a competitive environment, businesses must create a new space for themselves, by being the strongest in their chosen territories, whether that means product types or particular geographical markets—because if you're not the strongest in some area, it's simply impossible to survive. The important point is that before entering a new business territory, planners must ensure that they have a chance to win, and not overestimate themselves nor underestimate the strength of their challengers. Since the business has chosen to fight this battle, nothing but victory is acceptable; otherwise, it should withdraw.

Of course, if the territory won is bountiful, businesses have unlimited opportunities; but if unfortunately it's barren ground, taking it over has no benefit. Economists have spoken of the principle "winner takes all," but the question to ask is, "All of what?" Is it all of something negative, or all of nothing? Packard Bell got stuck with taking all of something negative because the personal computer (PC) market is one characterized by heavy sacrifice. So what is the use of taking all? To be a true winner means being a profitable winner; if all you've won is a burden, all that's to be done is to discard the acquisition.

A BORDERLESS MARKET ECONOMY

Finished products, like PCs, and the technology that they incorporate, like networking standards, all circulate globally, and this is the same for capital. Why is the market increasingly borderless? Because a growing proportion of items that everyone wants is not physical goods, which makes them easy to share. All that's needed is to create value, and even if the benefited party pays an extremely small fee, thanks to the large volume of sharing, it can add up to an enormous success. The vigorous development of the U.S. economy during the 1990s occurred thanks to technology sharing, especially in the area of software. Software that can be duplicated does not use up resources and costs almost nothing, but when software is shared, the utility produced is the highest.

Software cries out for users in the same way that books need readers—once they establish an audience that responds favorably, the profits will follow naturally. Stress needs to be placed on intellectual property, which can be shared on a large scale, making its added value high. Service expertise is also very valuable because, like software, the whole world can use it. On a case-by-case basis, service expertise is difficult and expensive to provide, but if it can be deployed over the Internet, it can be more easily duplicated and shared, and only through sharing can there be added value.

In the borderless market, for a concept to become globalized, local environments must be studied and understood. In the same manner, if a product is not the result of combining the best that the world has to offer, it is not competitive, and its success cannot be sustained. To effectively integrate the world's best requires global partnerships, and this implies that the future will not be an era where "if you're not my friend, you're my enemy"; rather, "dancing with the enemy" will be a common practice. In this environment of cooperation and competition, the ability to be open, flexible, and play different roles when needed will become important.

Recently, the notion of "decentralized organizations" has been much discussed. In fact, the term "decentralized" in this context is somewhat misleading as there must always be a "center" for any task, to initiate and lead. However, in a decentralized organization the "center" is not fixed, but is determined according to the particular task at hand. The same principle applies to entire industries—even if a company is small it can be the center at certain junctures, and the role of a large company may be that of a satellite supporting the center. This is the notion of a "virtual dream team," which will be discussed in depth later in this book.

Development of super-disintegration

The Internet era is one in which businesses will follow the super-disintegration model. Things that a company would have performed itself in the past are now carried out through strategic outsourcing as much as is possible. Why has this trend emerged? A business is concerned with performing a series of tasks, and of these only a particular few are core steps. A key part of strategizing is deciding which parts one should do oneself and which parts should be outsourced.

In the early 1990s, the U.S. began adopting this concept, which meant concentrating on core abilities and handing other tasks to outside companies. The advantage of outsourcing is that a company can stay focused, and even if costs are not lowered, the outside company providing the outsourced services is certain to be more expert in those areas—any lowered costs would just be an added bonus. The carrying out of in-house operations that are outside the scope of a company's core competence is inefficient because it requires personnel development, and during the process of training, such people contribute little and are basically idle resources. Even when the training is finally complete, it is often found that such personnel still lack a professional level of skills.

Super-disintegration works this way in the information technology industry. Dell and IBM were not proficient at manufacturing, so they outsourced it: insisting on doing it in-house would only have made it a burden. The September 1991 issue of the *Harvard Business Review* contained an article that is one of the conceptual pillars for the U.S. sustained productivity growth. The article maintained that the U.S. should have an economy with "computer companies that don't make computers, and fab-less semiconductor companies." This outlines the idea

Figure 1.1 Industry paradigm shift

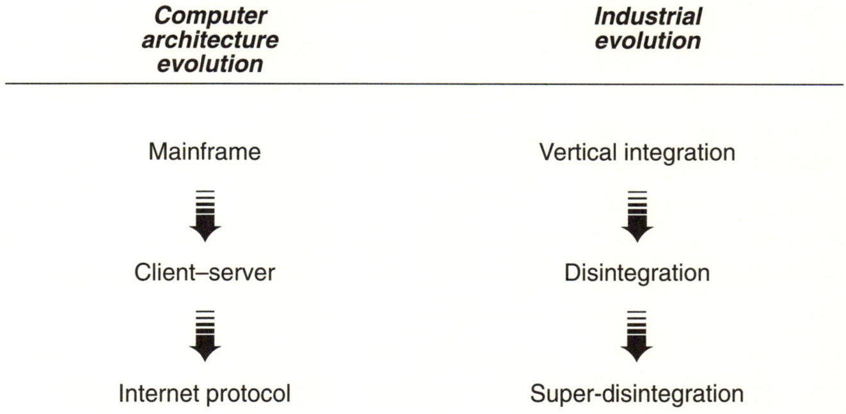

Computer architecture evolution	Industrial evolution
Mainframe	Vertical integration
⬇	⬇
Client–server	Disintegration
⬇	⬇
Internet protocol	Super-disintegration

© 2000 Aspire Academy. All rights reserved.

of outsourcing, which was quite novel at the time, even if it seems like an inevitable development now.

The basic precondition for super-disintegration is open standards that push everyone in the same direction by providing a common source of benefit for all, as well as providing a framework for continual competition. In the future, the notion of virtual teams and virtual integration will emerge. A team will be organized for a specific task, in an environment for cooperation provided by open standards and common interests to complete the task, and then disbanded. The U.S. basketball Dream Team was an example of this principle put into practice—it was a temporary team charged with the task of winning the Olympic gold medal. The best players were gathered together, and then went their separate ways after the gold medal had been won. Each player had a specific role, but everyone was integrated as a team, reducing burdens for individuals and enabling completion of the mission. In the networked world, this mode of doing things is more appropriate.

The development of computers can be schematized as shown in Figure 1.1 above. Network protocols are very simple things in themselves, but with the global open standards and self-governing system they make possible, the way the whole world works can be said to depend on them. Managing people and managing computer development have similarities, which can be seen from the parallel between the history of computer designs and the evolution of organizational structures. Computer architectures are continually being adjusted to meet changing requirements, moving from mainframe computers all the way to the Internet. The management of people and organizational structures will move in similar directions in the future, going from the ideal of vertical integration, wherein one does everything on one's own, to disintegration, and then on to super-disintegration. The difference is that organizations must focus on people, who can't be captured by the simple model of a computer.

As mentioned above, the future will certainly be an era in which relationships in business are at once competitive and cooperative. No single person, country, or company will be able to do everything. Without the ability to divide tasks cooperatively, competitiveness will also be compromised. As the saying goes, help comes to those who help themselves, and one must possess competitive advantages, have core strengths, and continually develop them, as well as create strategic alliances with partners to build a dream team. Under these circumstances, the willingness of many Asian small and medium-size businesses to be the leader in a market or product niche, rather than struggle futilely in businesses where they don't have the resources to dominate, is quite advantageous. That is because Asia's small and medium-size businesses are extremely flexible and willing to take on anything in order to survive. This practice is the most advantageous in the Internet era. During the last period of strong growth in the PC industry, relatively small Taiwanese companies triumphed over large Japanese and Korean conglomerates because of this way of thinking. We'll have to wait and see how to win in the Internet era.

A SHIFT FROM PRODUCT-CENTRIC TO CUSTOMER-CENTRIC

In earlier times, all industries could be sustained merely by having products and using technology to continuously develop additional functionality for them; this was the product-centric era. However, the future will employ a customer-centric development model, and even more importantly, many products will be oversupplied. Even with dynamic random access memory (DRAM), which represents an investment of billions of dollars, supply may outstrip demand. Relying on products and technology by themselves is not enough to sustain permanent growth and is therefore of limited value. No matter how many people you have, the labor resources of China and South-east Asia are far greater; how can you compete against that? Moreover, many things, especially intangibles, do not require physical labor.

In the future, we will develop more products, especially software, and if these products are not developed from a customer-centric mindset, they won't be wanted even if they're given away free. But on the other hand, if you've got a hot product, the returns will be unlimited. Not being customer-centric will inevitably lead a company to serious problems. In the past, the competitiveness of some of Taiwan's products was based on their being made a little better and being a little less expensive than others, but in the future, thought will need to be given to the question of what consumers really want. Do they want peace of mind and convenience? Or is it a low total cost of ownership, or a low price for just the product itself that they prefer?

Intel is a good example of a company that is very customer-focused. Intel's products are manufactured by thousands of OEMs (original equipment manufacturers), but its true end-users are the tens of millions to hundreds of millions of people who use PCs. Intel aims to exploit the customer-centric approach, and its design focus is on meeting actual customer needs. For example, customers want to be able to upgrade at will and have software compatibility, so Intel sells them on the idea that "buying the latest technology protects you from obsolescence." In its advertising, it stresses the same message. Intel does not utilize an intermediary, such as Acer, to communicate this idea, because the two companies have different objectives. So Intel skips Acer and communicates this information directly to consumers.

In view of these global trends, a business should ask what its positioning is. Is it a company whose primary mission is to provide products, offer services, or supply solutions? It's important to establish company positioning, especially in this era of industry disintegration. It can be likened to playing a team sport—every member of the team must think about where he or she has to stand in order to shine, and this will be a deciding factor in whether the team wins.

Businesses must understand customer requirements. Taiwanese companies, especially those in export industries, are heavily dependent on OEM and ODM (original design manu-facturing) business, which simplifies things. Creating one's own brand is not as simple because thought must be given to who one's customers are and what they want. Inattention to this issue is why leading companies, such as IBM, with powerful brands, gradually lost those customers who had less money to spend, and in the end even the most moneyed clients did not necessarily buy IBM products because its products and those of its competitors were not very different. If you don't think about what your customers are thinking about, you'll eventually run out of room to expand your business. Therefore, for companies to create their own brands, they need to get close to their customers and understand them, and then create a workable business strategy step by step.

In trying to become customer-centric, the main obstacle for companies is employee mindset and corporate culture. For example, Taiwan's industry is export-oriented, so it is distant from consumers and thus unable to understand consumer needs. The attitude of employees is

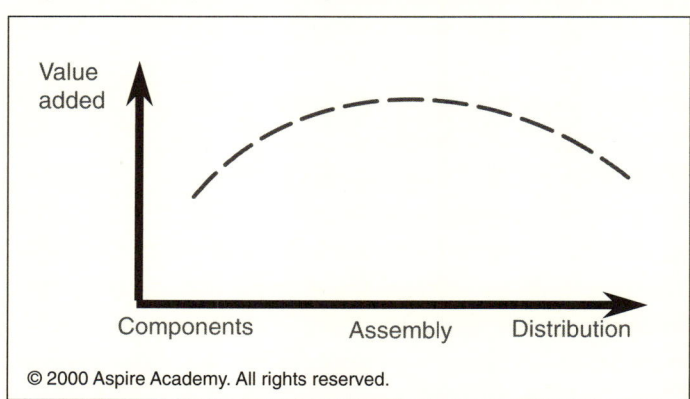

Figure 1.2 Computer industry: value-added curve before 1980

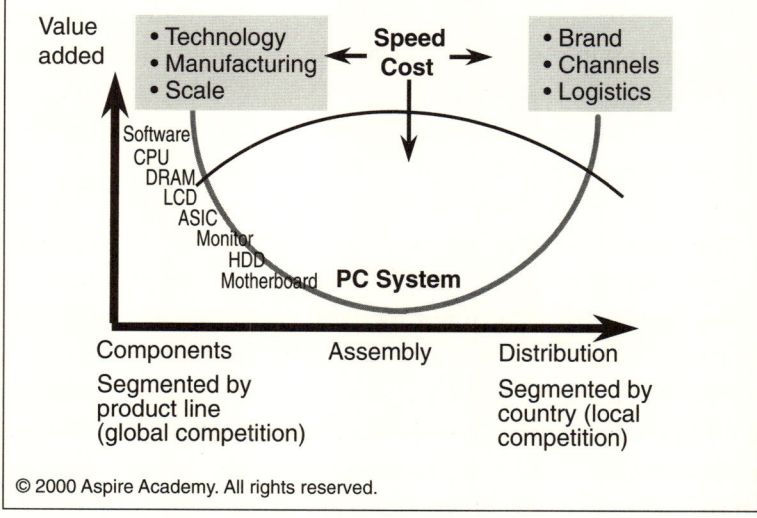

Figure 1.3 Stan Shih's smiling curve—sources of value in the PC industry (1)

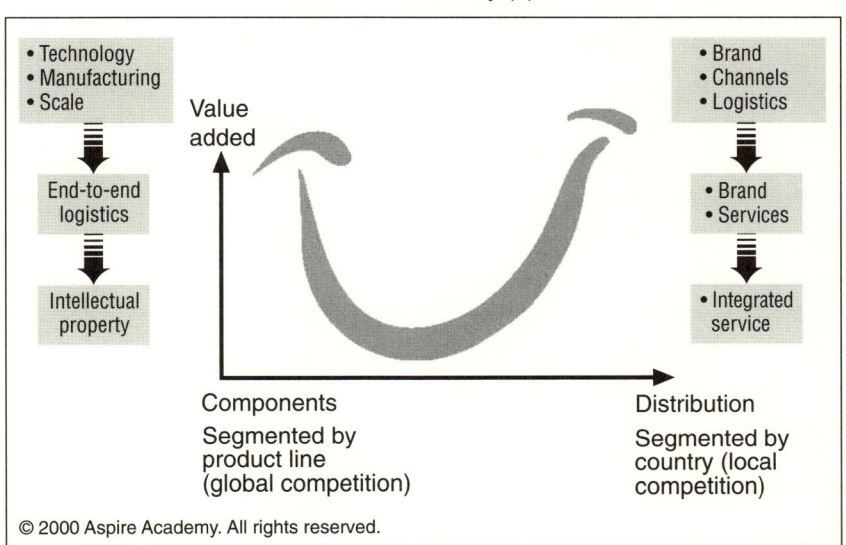

Figure 1.4 Stan Shih's smiling curve — sources of value in the PC industry (2)

that they should do whatever the largest clients want, and they don't understand how to listen to consumers. Transforming this attitude into a truly customer-centric ideal is fraught with difficulties. Another factor is the nature of the industry; for example, PCs are product-oriented. In the PC niche, Acer has its own brand, but the product direction has already been set by the Wintel standard. When Acer tried to innovate, it failed, whereas companies that just blindly followed Wintel succeeded. Innovation became a burden, meaning extra effort without any returns. Some U.S. companies, such as Hewlett-Packard (HP) and Wang, also wanted to innovate, but in the end gave up and lost out to companies that just went along with Wintel.

A SHIFT IN THE SOURCES OF VALUE CREATION

Where is value really located in the value chain? In the past, mass production was the key, but in considering the future, intellectual property that is used on a large scale and services that customers access directly will be more valuable. A simple plot of how added value works makes this clearer (see Figures 1.2 to 1.4).

Though the figures are based on concepts relevant to the PC industry, they also apply to other industries. All industries can consider where value is actually located in the value chain that begins with initial research and development (R&D) and components, and ends with marketing. It is obvious that prior to the 1980s, the greatest value in the computer industry came from assembling finished products (see Figure 1.2); however, by the early 1990s, this curve had been flattened out, taking the form called Stan Shih's "smiling curve" (see Figures 1.3 and 1.4).

This smiling curve originated during Acer's re-engineering effort in 1992. How could thousands of Acer employees be persuaded that the things we had laid stress on in the past were no longer as worthwhile? To do this, I drew this curve and passed it around. After the idea had been disseminated, everyone joked, "If you don't understand this curve, you can't smile." Clearly, the added value of manufacturing has already diminished.

In the area of basic technology and components, competition must be seen as global, but in the area of sales and marketing, the competition is localized. (This simple concept will be reiterated later in the book.) If this direction is misjudged, and business creation and development violate this principle, there will be problems. In the smiling curve, the locus of important added value is always changing, and the value created by whoever possesses more customer-centric intellectual property will be the highest.

Looking at the situation today, all indications clearly point to software as the locus for the highest concentration of such customer-centric intellectual property. Bill Gates once told me that he didn't believe Intel was just a semiconductor company; he thought of Intel as a software company and as Microsoft's greatest competitor. To be sure, Intel is an intellectual property company, and all it does is integrate its intellectual property into silicon chips. Anyone can make chips, as some parts of Asia can and do, but the key is what is inside the chips. So Intel does not only make chips, but the chip is a container for its intellectual property, and Intel's semiconductor business is in fact the selling of intellectual property.

Under these circumstances, we cannot help but ask: "Having worked so long, have we actually created added value?" In the larger disintegrated business environment, if you can't be the market leader, does it mean you should give up? Are you willing to dominate a niche if it means playing a lesser role in broader segments? Acer sees it this way. We specialize in creating new companies that excel in their particular line of business. The idea is that if you can't be sure of doing a good job, don't bother. However, if you find a market that has value for the future, you must put everything you can into it and become the leader in this new market.

The entrepreneurial spirit of Taiwan's small and medium-size businesses has two elements: the first is the willingness to be a niche leader, and the other is a "me-too" mentality. The former should be encouraged, but not the latter because it leads many companies to pursue the same niche. Look at the exit of Intel as well as Acer and other vendors from the DRAM market. The reason for their action was that there was no way for these companies to take a large share of that market, so the only reasonable alternative was to pack it up.

THE DIGITAL REVOLUTION IN THE INFORMATION ERA

The digital revolution is creating the New Economy; and in various places around the world, especially the Middle East, South-east Asia, and Latin America, it is possible to see the impact of changes brought about by the New Economy and also understand that the U.S. is playing the key role. Even though people in these places can keep up with the latest trends, they unfortunately have no role to play in its continuing development. At present, they are only spectators, and we hope that North-east Asia can avoid such a fate and play a role in the present scene. In the new economic order of the Internet era, we must be actively engaged in obtaining a role; perhaps it won't be the starring role, but at least it should be a key supporting part.

In the digital revolution era, customers and companies will no longer obtain market information at widely separated times. Information is so plentiful and accessible on the Internet, and customers are extremely attentive to market conditions. Meanwhile, company employees may only understand what's required by their work and not seek out various types of knowledge as actively as consumers do. If employees do not actively seek expertise in areas related to their company's business, the unfortunate result may be that customers are more knowledgeable than the companies.

In the New Economy, things are happening everywhere, and every day we're in the situation of having to react without preparing as fully as we would if there were more time. What is the right approach to take? The critical factor is not so much the amount of raw effort put into the preparation, but how well we can utilize resources, move in the correct direction, and lead our organization to do the right things.

CONCLUSIONS

Only through evolution can success be easily achieved; adjusting to truly revolutionary changes, on the other hand, is very difficult. Is digitalization a revolution or an evolutionary process? For someone who doesn't think about this question in any depth, digitalization is a revolution; for someone who observes attentively, it's a step-by-step evolution. Understood as a process, this evolution even allows you to understand in advance what developments the future will bring. We must always keep tabs on the actual environment and then move forward aggressively.

When problems develop, the means of change must be sought immediately, first getting a grasp on the situation and then making reforms—starting with yourself. Moreover, when a problem is discovered, it is often quite late, and if you want to look for excuses and not reflect on where you can improve, it will just delay the improvement process. From this, we can see that the re-engineering of Acer actually took place a little late. We saw that the company had problems, and even though we did some self-evaluation, in 1991 we started seeing losses without even a chance to look for excuses. However, we didn't start the re-engineering process until 1992. In 1996 and 1997, when we made another re-engineering effort, profits were already declining by the time we started doing some soul-searching. The record shows again and again that Acer's ability to look critically at itself was lacking.

After pressures accumulate for a long period of time, a turning point will be elicited, and its unfolding is quite abrupt. Perhaps you don't have a grasp on what a turning point will be, but still you must be prepared for it. When the turning point arrives, you must be able to take advantage of the opportunity to make changes and adapt quickly. However, you must also avoid jumping in too early, throwing in all your resources before the turning point has truly developed. In that case, if you spend too many resources, you may exhaust yourself before the race is over. For many people, the Internet is such a turning point; the growth of Linux may be a turning point for the industry dependent on Windows, for which you must be prepared. Even more importantly, you must establish an appropriate organizational structure and culture, as this will make it easier to adapt to and even take advantage of the change.

DISCUSSION

Q: In the course of operations, business executives make decisions and utilize internal communications to obtain a final consensus. But how does one choose between thoroughly considered policy and decision-making efficiency?

A: You must assess what exactly you will be able to get in return for spending so much time in communicating. If the sacrifice of this time is not detrimental, but it can lead to more people getting a full understanding, then it's worth doing. If a similar situation arises in the future, and the organization has already cultivated these communication channels and can reuse this capability, then this decision-making process is an added value for the organization.

Even more importantly, the people in charge of executing decisions are not organizational leaders, who must use a quick and decisive mode of thinking to lead everyone in the same direction. If you're not careful, what ends up happening is that the people higher up say one thing needs to be done, but the people under them don't know how to carry it out. I'm very patient in this regard because I use time to buy experience, which is an investment for the future. I invest time today to communicate within the organization in the hope of saving time in the future.

Overcoming difficulties as a means of creating value

Q: Based on several decades of running a business, do you believe there is one theoretical framework that can cover all business models?

A: My initial way of thinking about this was that you should deliberately pick a difficult task to do, and that by overcoming obstacles, you can create value. The whole basis of my business was in this idea. It didn't mean not doing anything simple, but it did mean looking for difficult things to make a breakthrough with and create value—this was the overall framework. While you're dealing with obstacles, you can't forget about taking care of the routine simple things, but if in running a business, you only choose to do the things that everyone can do, what's the point?

The second mode I considered was that you had to have a vision—identifying objective factors and understanding the external environment, analyzing strengths and weaknesses, and protecting weak spots. This should be the starting point for developing business strategies. For example, when Acer first started, capital was insufficient and staffing was short as well, so I created a plan for both finances and personnel that took these weaknesses into account.

Can Acer's hardware business become the world leader? The opportunity might exist, but preparations have to be made, which is what I'm talking about: you don't know when the turning point will arrive, but when it does, are you prepared? Of course, Acer is preparing itself step by step.

As for the software industry, the influence of the market is even stronger. The link between software and the market is even deeper than with hardware, so if the software you produce does not have a robust market, finding room for growth is difficult. Therefore, beginning in 1997, Acer has held software strategy meetings and decided on three directions: integrating software with hardware, developing localized content, and implementing localized service.

In the disintegrated business environment, where the market is sliced up into small pieces and you supply particular markets among them, you have to be a leader in whatever you do. After cultivating your capabilities, you start working toward the next objective. This means facing the biggest problem: what you can count on getting may be just too limited, an individual market is too small, and it takes so much to develop the proper scale. And yet in the New Economy, you have to be focused. So the question becomes: how can you be diversified yet focused?

In Asia, business diversity in the past was achieved through protectionism, the dominance of financial conglomerates, and collusion between government and business; in the diverse business environment of the new, freer economy, the question is: can you stay focused competitively? What's the use of being big? If you don't excel at what you do, how can you compete?

Diversified yet focused

Q: **Some people say that the Internet has narrowed the distance between vendors and consumers, making brand management very** important. Many companies in Asia are focused on OEM business, so in the New Economy, what is your advice for these companies?

A: There are many products and technologies that are globalized with a unified standard. But if you want to enter a market, you must have something very localized. Customer requirements and service are very localized, and that's something that won't change. However, Internet technologies will change and enable businesses to go global very rapidly; if there were a set of methods to get companies localized quickly, that would make globalization much easier.

In the New Economy, it's difficult for a business to do everything on its own. If in a certain area, the Acer Group aimed to do this, then a diversified focus would be called for, meaning that at every phase it would have an independent focus. A single company or business unit cannot do everything on its own—there's no way to compete that way. In the future, the key will be how to effectively cooperate with local partners.

The important point for effective cooperation is to give precedence to markets that are close by. The advantage of being close is not only making understanding these markets easier, but also that you get very quick feedback on what you're doing. For the same type of situation, you can rehearse a few times and then you can handle it easily.

Brand management is also very important. The value of a brand is accumulated gradually, bit by bit, and the main goal in creating brand image is that it is advantageous for winning repeat business and ensuring your business has continuity into the future.

The boss is the biggest obstacle to re-engineering

Q: **Super-disintegration is a trend in business, but after the products become more and more integrated, will there be a conflict in the market?**

A: Integration is just one type of subdivided task because after the opening of standards, integration became much simpler. Sometimes consumers play the role of integrator, and they choose for themselves the best parts available on the market and perform the integration. The key point is that with proprietary systems, nobody understood integration. For example, take the sale of early computer systems to Taiwan. In the case of IBM, they had to send a few people to Taiwan to live and train the staff because only a few people understood the system. After the establishment of standards, many people are able to perform different kinds of integration services that customers need.

Q: **You felt that Acer was a little late the two times it undertook to re-engineer itself. At what time do you believe that a business should undertake reforms?**

A: Re-engineering know-how must be developed at every level. Taking Intel as an example, when top-level management discusses changes, the people under them have already been thinking along the same lines. The top level is the slowest. Unfortunately, all re-engineering efforts have to be initiated at the behest of the people at the top. So if the re-engineering consciousness is present everywhere, this information can be passed very quickly up the organization, and the decisions can be made quickly, and re-engineering may even start taking place without a conscious decision.

An organization often needs to be re-engineered because it has gone too long without improving itself. If the people at the lower levels of the organization are constantly doing small-scale re-engineering tasks, then the organization as a whole won't need to undertake re-engineering. The question is whether it's the executives who need to be skilled at re-engineering or the organization as a whole that needs to have the ability to re-engineer. This might be the key ongoing issue for businesses. In general, the biggest obstacle to re-engineering is the person at the top. In the U.S., corporate re-engineering means changing CEOs; without that, how can you re-engineer?

Stick to what you're good at

Q: **Many companies in traditional industries, such as construction or heavy machinery, have started up Internet businesses, as Acer has done. For companies in different industries to evolve into Internet businesses, what different methods are needed?**

A: In a protectionist environment, the problem of over-diversification often arises. Asian companies often put a lot of stock in being jacks-of-all-trades, seemingly able to do anything. In a democratic society, people are particular about acquiring specific expertise, and it's difficult to be everything to everyone because every field has its own culture and specialities. Even if you aim to be a jack-of-all-trades, you cannot count on being able to always use the same old bag of tricks, but you must focus your efforts on particular areas, such as building corporate or brand image, accumulating capital, or developing management expertise.

Looking at it in another way, in new areas of business there will inevitably be novel demands created by the new environment. The key is whether there is enough time to establish a new core competitiveness. In Korea and Japan, the pie is always being divided up among the same business groups and financial conglomerates, regardless of whether they do the job well or not. Because the market is shared by just these few companies, they can do whatever they want. This way of doing things just doesn't work anymore—to be a successful player in the New Economy, you need to perform well, and you've got to have a well-thought-out approach.

Acer's idea is not to diversify outside its area of expertise, but to confine itself to the information technology industry; or if it does go outside the industry, the area must be related to the original business. Many people have invited me to be a bank president, but I was very forthright in turning them down for two reasons: first, to not go outside the area of my expertise and, second, to avoid a confusion of roles because a businessperson and a banker use a different base of knowledge and play different roles.

Regardless of what a corporate culture or organizational structure is, you have to fit the role you play and make sure that you don't win through size but through focus. If you don't have this determination and just want to hold your ground, you won't survive.

Q: In the New Economy, what qualities must a successful business leader possess?

A: The first is boldness. The second is having your own point of view—if your point-of-view is just "me too," the people you're leading will not be satisfied. The third is knowing how to take advantage of people's talents; communication is very important.

CHAPTER 2

Using Small Size to Win Big—
The route to success in the Internet economy

THE LOGIC OF THE INTERNET ECONOMY

IN THE INTERNET AGE, businesses will not compete on the basis of who has more people or covers the bigger territory, but on the value they create. The most important consideration in the era of the Internet economy will be whether you can create valuable "bits"—digital information, content, or solutions.

The most distinctive feature of the Internet is that avenues for doing business are everywhere. The Internet has stimulated the rise of the knowledge-based New Economy. In the past, knowledge would have been exchanged face to face or through print, and its results would have been transmitted via some physical medium, compromising its effectiveness. Knowledge in the Internet economy is disseminated as bits, and the influence it produces may reach almost anywhere, challenging old traditions and values. In the New Economy, the way to wealth has been drastically altered. The mode for creating wealth advocated in the past, even if a few principles still apply, has been affected by the advent of the Internet in such a way that it may no longer be of much use. One of the precepts of the Acer Group's SoftVision 2010 concept is creating "human-touch bits," with the objective of creating valuable, commercializable digital information, content, or solutions—this is the most crucial consideration in the Internet era.

What is a "Web business"?

The Web business is a low-tech business. By all means, do not see the Internet as high-tech; in fact, it's no-tech. I am not denying that the Internet is based on advanced technology, but with the minute division of labor that goes into making it what it is, most of the business that can be done does not require technology. However, for something free of high-tech to produce value, it must possess "high touch." Despite lacking technology, if something has this high touch, it can increase value.

The Internet is a new business opportunity, but all older industries can use it to re-create themselves in a new form. Furthermore, one should not assume that at present, there are leaders in the Internet economy. Even if there are a handful of people doing a lot of moving and shaking, it's no big deal. Right now, everybody is usually doing the same things because the barriers to entry are minimal. To see what can really be done, more time is needed. The most important things are whether the market segment that a company chooses possesses enough business potential, whether the company can innovate, and whether it can execute such innovations.

In the traditional economy, getting a business to the critical point in terms of scope required more resources, but things are different in the Web economy. Right now, it's not that important to have new ideas; the true value comes from being able to execute ideas effectively. The Internet economy is knowledge-oriented, and this knowledge is not just erudition but includes much domain knowledge.

Where is the value or the returns to be found on the Internet? Many people see the Web as a new media format that depends on advertising for revenues; there are people who see the Web as a sales channel, as a marketing avenue for business, which gives rise to e-commerce; others take the Web to be a type of public infrastructure, like water and electricity utilities—you have to pay for what you use. The Internet is also a new type of community, and to serve this community, the members must pay membership fees. The Web can also be seen as a type of technology platform, a data center for security and financial transactions; during the process of selling technology, everyone follows the super-disintegration model and shares the costs. A Web business can also be a type of consulting and query-answering business, where the returns come from collecting service fees.

The challenge of Web business

To completely transform the old economy into the New Economy, the biggest bottleneck to be cleared is the shortage of people; this is a problem that the whole world faces. From the perspective of smaller countries, such as many of those in Asia, the challenge of doing Internet business is that their local markets are too limited and the economic scale is too small. Especially in the case of knowledge-based industries, such as the Web industry, the larger the economic scale, the more effective it can be. Countries have no choice but to get involved in the Internet economy, but if they don't look outward, the discrepancy in what they can achieve, compared with other countries, will be large.

This situation is quite different from that of the hardware business. Semiconductors made in Asia can be sold throughout the world, but can Web content produced, say, in Taiwan be propagated globally? How can the Web services provided in one part of Asia be extended to reach overseas customers? Up to the present, the barriers to entry for the Internet economy have been minimal, and innovation has been very limited as well. The Web as it currently exists seems to change its nature every day, but this represents a new opportunity. There is a lag between the time when people perceive an opportunity and the time when it has been fully exploited, so people will try to take advantage of the opportunity ahead of other people. But without innovation, by only repeating what others have already done, it's open to question whether they can survive for long.

In addition, as business models for the Internet are still being formulated, assessing the desirability of entering a new area is a big challenge. Though the U.S. is in the middle of an evaluation process, everyone is of a different mind about it. The experts in the banks and in industry feel they're in the dark, to say nothing of everybody else. There are many people in the U.S. who are taking a gambler's stance, and all that's needed is for these people to place their bets and the money starts to grow. What needs to be noted is that the U.S. is a large economy, and the same thing done there and in a smaller economy will produce very different results. Judgments made in the U.S. about the Internet can't necessarily be applied elsewhere. Hong Kong, which is caught up in a Web business craze, has discovered that there is a lot being done, but there is no market for it the turnover is little.

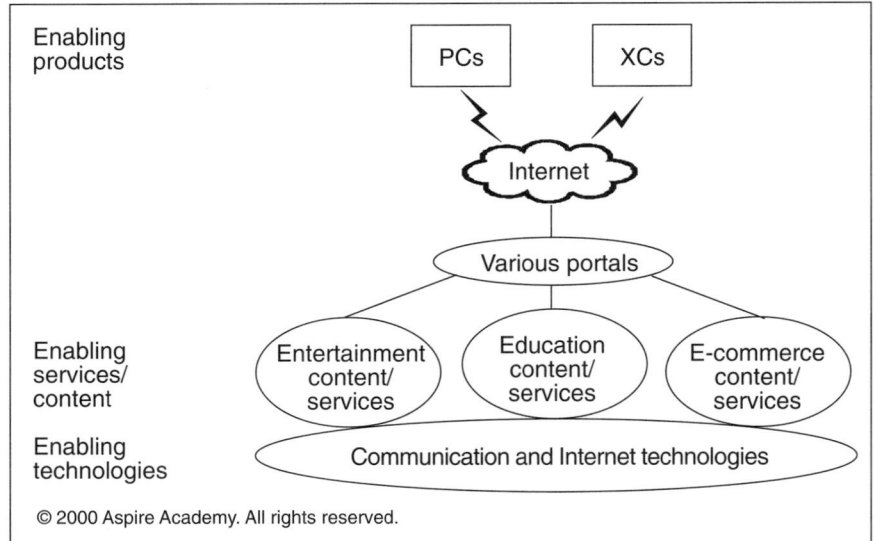

Figure 2.1 Enabling the Internet economy

In making the New Economy a reality, there are three key areas to consider (see Figure 2.1): the first is to stimulate its active development by providing the tools and products that people need to get involved in the economy (enabling products), and make their use as simple as using a telephone. At earlier stages, the computer is the enabling product, to be followed by single-purpose computers (X-computers or XCs) and information appliances (IAs) or cellular telephones.

This first area is relatively simple for Asian companies to deal with. A second area to focus on is services and content that enable development of the Web, and these must be convenient and useful for consumers to access. Some mechanisms must be created that involve various types of portal sites, and these are divided into the "3E" categories—e-commerce, education, and entertainment. The third area is nurturing the Internet economy with enabling technologies. The technology that underlies much software and communication, and even security and financial transaction mechanisms on the Web, can expedite the spread of Web services and designs.

ASIA'S STRATEGY FOR DEVELOPING E-TECHNOLOGIES

In the area of products, Asia can use the advantage it possesses in the manufacture of computers and related components to become the world's supply center for e-products, and this is the comparatively easy part of the task. Using this approach can lead to large volumes, but the potential for adding value is limited. Another approach that offers greater opportunity to add value is to take advantage of Asia's semiconductor manufacturing prowess and develop information appliances and single-purpose computing devices, as well as to do R&D for the software embedded in system-on-a-chip solutions. By supporting the development of the Internet economy, these products create added value.

In the area of technology and the products based on them, these innovations must be globalized, while services must be localized. The technologies used must be the most advanced, but that does not imply that you have to develop all the technologies yourself because the cost of doing so would be too high. Even if you did develop them, they might turn out to be inferior to what your competitor developed. In technology, compatibility has to be considered because the Internet is an interconnected medium, and whether a technology can win acceptance worldwide or not is very important.

Therefore, in the area of technology, it's essential to work with global leaders or obtain licenses so that the core of your products is true leading-edge technology. Alternatively, if you develop technology yourself, there is a chance to become a world leader in some market segment because the scope of the Internet is enormous. A good example of this is Chinese-language search engines, where Taiwan has an opportunity to become the world leader. To succeed in a technological area, companies must stay focused and quickly adopt global industry standards.

ASIA'S STRATEGY FOR DEVELOPING E-SERVICES

Since direct contact must be made with customers, Web services must be localized. The use of the Internet is a global trend, and to get into Web services, less well-established companies can partner with big multinationals to take advantage of their technology and experience, as well as their brand recognition.

In the future, competition to establish brand advantage will be even more intense in the Web economy than in the traditional economy. There are two reasons: first, there are too many brands on the Web; the field is too crowded, and hence establishing a positive brand image is even more important. The second reason is that the Web deals in intangible items.

Credit is critical for doing e-commerce, and having an established brand carries an advantage.

In Web services, customer relations is most important; the informational content provided by the services must be relevant for local users. Internet technologies should be uniform across the globe, but if services cannot meet localized needs, it lacks competitiveness.

INTERNET BUSINESS OPPORTUNITIES

The Internet provides a wealth of new business opportunities, and these are open to everyone. Even though at present there isn't much sign of innovation, with the trend toward super-disintegration, in the future, business and its methods should provide ample space for innovation. In fact, without innovation, no value is created. Something we should be grateful for is that Web services must be localized; if you can become the kingpin in the local market, you have a big advantage.

In order to survive in the long term in the Internet environment, a relatively small amount of resources is needed. The situation can be compared to establishing a factory—it must have at least a minimum scope, and that implies the use of a sizable amount of resources. A roadside stall or a small shop doesn't require nearly that amount of resources, and yet many actually have a longer life span than a factory. Should a business be run with the aim of remaining small but for a long time, or should it try to hit it big and go out in a blaze of glory?—that is a matter of personal choice. The super-disintegration of the Internet business creates a very fragmented market, which means there's room for even a single person to make a go of it.

How does super-disintegration in the Web economy differ from the traditional division of labor? A production line is also a form of division of labor, but each division in the process cannot stand independently; disintegration, on the other hand, means an effective mode of operation with the value chain split into many segments, each segment representing a stand-alone line of business that can be sustained independently. Division of labor in the past meant "upstream," "downstream," "customers," and "suppliers." These referred to fixed entities; however, under the super-disintegration regime, their identities are no longer fixed for each segment, and all are independent of each other. This disintegration model is extremely common in the Internet economy. In addition, because the

Internet is not a physical entity, mass production doesn't entail high costs. So if you get things right, the net gain will be great.

CREATING VALUE IN THE INTERNET ECONOMY

In the Internet economy, how is value produced? Most people focus their attention on the fact that the value of B2B (business-to-business) transactions is potentially very large (roughly in the range of tens of billions of dollars or more) while B2C (business-to-consumer) transaction values will likely be very small in comparison, with little profit to earn, leading to the conclusion that effort should be focused on B2B. However, in gauging value in the Internet economy, we should look at the fees collected from e-commerce transactions, and not the total value of the transactions themselves. All the reports mention how B2B involves billions of dollars, and how much more this is than B2C, but is this really more sensible? The conventional wisdom of emphasizing transaction value and not transaction fees is misleading, and the conclusions reached therefore should be treated with skepticism.

In the past, doing business meant talking about how much market share you had, but in the Internet economy, the key thing to look at is the proportion of the total value that you create. Another misconception is that in the Internet economy a winner can beat out all other competitors and completely take over. The real question to ask is: take over what? If you take over something negative, it's your loss. If you take over something meaningless, all your effort was for nothing. Furthermore, the advantage of the Internet is that even if you get beaten, you can still make another go of it. Right now, everyone is trying to do portals, but this approach is open to question. Once again, the Internet is about super-disintegration; there is an unlimited number of market niches in which value can be created.

The Internet has another area with great potential: companies can use Internet business models and technologies to lower costs. Creating a new market is not easy, but companies can use the Internet to reduce their present costs of doing business. The value created in this way is not less than that from new business opportunities. In thinking about the new Internet economy, my competitiveness formula is also applicable (for details, see Chapter 5).

A POSSIBLE BUSINESS MODEL FOR AN ASIAN INTERNET INDUSTRY

A workable business model for the Internet industry in Asian countries has not appeared yet, but it will almost certainly reflect unique conditions and turn out quite differently from the Internet industry in, say, the U.S. In the past, the successful approach that Taiwan, for example, used to take advantage of its two biggest business opportunities—one being computers and the other being semiconductors—was completely different from that used in the U.S. Moreover, after success in the U.S., Taiwan followed several years later using a different approach. For example, around 1985, Computer Land was the most successful and the best in the world, while now in Taiwan, Synnex International is the strongest. The U.S. has many computer and semiconductor companies that have established themselves globally, after several years of growth, and their competitive strengths are naturally different from that of Taiwan's businesses. Acer's competitive strengths, for example, are different from Dell's.

In the OEM and ODM business environment that many Asian companies are familiar with, each industry has five to ten players because foreign OEM customers look for several potential partners, making a "me-too" strategy viable. In the Internet industry, though, the winner takes all in its particular niche, and the room for "me-too" players is consequently much more limited. Doing Internet business requires building a brand, and this is also different from the OEM model. With computers and semiconductors, manufacturing capabilities have been the main competitive weapon, while for Web services, marketing is the key.

ESSENTIAL KNOWLEDGE FOR E-BUSINESSES

The Internet's potential as a business tool is as a medium where new business opportunities can be discovered. However, too many people see the Internet as a new business while neglecting its importance as a tool.

Not getting involved in Web-related business is fine, but you must not neglect the use of the Internet as a tool for enhancing your current operations. Therefore, seeing the Internet as a tool is vital. Many companies have added ".com" to their names and this may have a beneficial psychological effect. Many companies in Hong Kong have done this, with even real estate companies becoming dot.coms; however, just changing a name is undoubtedly not enough.

Companies investing in the Web will have no problems if they think of it as a tool. However, if they think of the Web as the company's new business, there is a great risk attached. You don't have to be a slave to the latest trends, unless you really understand the nature of the business you're getting into or have a real need to do so.

Markets in the Internet economy are still very small, so any investments made must be carefully considered. If you spend freely to try and match the scope of the market, it's a losing proposition because without returns, even being at the cutting edge just means being the first to run aground. With a small market, the resources invested must be correspondingly small.

Conclusions

Investing in the Internet means having sufficient means, but without investing or getting involved, it's impossible to acquire the knowledge or technology; you have to stake out a position while remembering not to think of it as a must-win proposition. Web investments are also risky, and the ability to handle the burden must be considered. The Internet is a type of technology that stresses being global and global products. However, competitiveness in the Internet, if you're talking about Web services, means stressing localization. You should have unique competitive strengths in order to have the ability to develop over the long term.

On the Internet, with market segments so numerous, business models are not fixed. In the future, whether cooperative ventures or mergers or other modes of integration will predominate is still unclear. In the process of investing, there is only one thing that is a sure bet: when what you invest in is ahead of what someone else is investing in. The Internet is based on super-disintegration, and if you possess a lead in some small area that helps enable a new service on the Internet, then you have created value. When another entity is trying to come up with a successful business model and lacks the piece that you have, then your investment becomes valuable. Simply put, you have ensured your selection as a member of the dream team that someone else is putting together.

In the Internet age, mergers and acquisitions will become commonplace, and the key question will be whether the task you're devoted to has value. The central issue for business is developing core competitiveness, but if a business group wants to be big it must be diversified, so the balancing of focus and diversity is very important. In addition, although businesses must be focused, they must also understand a diverse market, the dynamism of the Internet, and the varied needs of different customers. That is to say, they have to be focused on their task, but still have a grasp on the diversity of the broader environment.

In the Web economy, everyone is after speed, but effective execution may well be even more important than raw speed. However, looking at it from the other side, creating value may be a melding of both the new and old, because traditional businesses must create new value, and the best way to do that is to work with the new generation of businesses. Traditional businesses will get a larger piece of the Web economy pie because their investment is larger.

In the U.S., there is much discussion these days about whether it's better for large corporations' B2B or B2C activities to be part of their internal operations, or to have independent companies handle them. Some believe that new companies should be created to focus on this area because it involves a different kind of culture, and it's more effective to avoid the influence of the established large corporation; or they may believe that the market value of new companies is higher. If a large corporation separates its original business and Web business into two companies, the combined value will be much higher than the value of the original corporation.

There are other people who maintain the opposite position, insisting that the real value created derives from the resources controlled by the established company, with the knowledge and skills needed to do business contained there, while the new company's only strength is in technology or conceptualization. In the long term, value creation still relies on the older company; the best way to go is to keep the Web business internal, as a department within it, rather than in an independent company, as this will be more advantageous for overall competitiveness. Spin off a new company and its market value is high; keep everything in-house and it's more effective, and the created value is higher.

The question of which to choose is a big headache for large corporations in the U.S. because in the early stages running things within the culture of a new company is better, but in the long term keeping value creation in the established company is more advantageous. However, if any conclusions can be drawn at this point, it would be that keeping B2C e-commerce in-house is more advantageous because it allows the company to better coordinate its overall sales strategy. On the other hand, because B2B e-commerce basically follows a supply chain model, an independent company fits in the better with the scheme. The decline of stock prices for Web companies has greatly reduced the enthusiasm for companies to spin off e-commerce subsidiaries.

Discussion

Setting the price of engineering expertise

Q: How can one assess the value of a Web company?

A: To evaluate a Web company, you must look at whether the areas it's focusing its efforts on will gradually develop into a source of competitiveness. Without a distinctive opportunity to establish a lead over competitors, and if there is a danger of its being usurped, then you have to think twice about whether the company is worth investing in.

This involves another important factor: that there will be a shortage of personnel resources for Web development. In the U.S., when a Web company is considered for purchase, the thing that's looked at is how many engineers it has. Then a price is calculated based on a headcount of these engineers, regardless of exactly what it is that the company does. Therefore, whether what these engineers are thinking about suits your purposes, whether you can communicate effectively with them, whether you share a common viewpoint—these are all more important than what the company they work for does.

Q: **Many Web companies are still spending huge amounts of money without any visible returns. How will the situation evolve in the future?**

A: There are still misconceptions when the Web economy is discussed; for example, that anything new is worth investing in, even if it loses money in the early stages. However, communications and other physical infrastructure businesses are definitely ahead of the Web services industry, and the industry viewed with the greatest optimism is still wireless communications, and extending this to the Internet is very simple.

Running a Web business requires vision

Q: So, what does a successful Web business require?

A: The first wave of Internet mania was sparked by advances in telecommunications technology. The Acer Group's HiTRUST supplies network security technology and is already a profitable business. The general public doesn't know this company, and there is no advertising for it—it just works behind the scenes. However, if you want to do e-commerce, you have to work with it.

When a company creates services, having large numbers of customers is beneficial, but the key point is the value you give to these customers. The nature of doing business on the Web is still not fixed, and as it continues to be extended, its potential benefits also increase. It may be that B2B only presents an opportunity in technology, with the real specialized expertise still controlled by large corporations.

B2B is concerned with transactions, while B2C is all about services; taking an existing business and moving it into the Web space is the most optimal way of doing things. Doing B2B doesn't require establishing a brand, while entering the B2C space means you have to think about branding and the customer's sense of security, as well as needing to create payment mechanisms. B2B doesn't present operational difficulties, but B2C means distribution problems. So, in doing business on the Web, you must have a vision, take action immediately on the most pressing issues, and get familiar with factors you are not certain about. The digital economy will arrive sooner or later, and no one will be able to stand outside; otherwise, you're out of the game.

Q: In the past, companies used patents to establish their core competitiveness; in the Internet era, how do companies establish a distinctive competitive strength?

A: The problem of intellectual property rights naturally will carry over into the Internet age, and it's related to patents for physical objects and copyrights for printed works; software can be patented. Competition in the future will take place more in an invisible realm rather than in a visible one. Intellectual property includes brands and customer relations management, and these can cause obstacles to competition. Further on, personnel resources can create competitive obstacles; personnel resources doesn't mean having many people, but having an organized team that can work together.

Marketing that reaches the Chinese-speaking market

Q: Marketing is a core competitive factor in the Internet economy. How should marketing be pursued for the Chinese-speaking market?

A: Marketing requires an understanding of the market and local cultures. Markets are dynamic, and effective marketing constantly monitors and coordinates with changes in the societies in which the markets are situated. In the most basic marketing experience and theory, the U.S. is the strongest, but to apply this expertise to other markets, integration with the local culture must be taken into account. In marketing for Chinese-speaking markets, Taiwan has a decisive advantage. The current problem is, who owns this market? Is it Acer in partnership with foreign firms? Acer alone? Or Acer working with mainland Chinese partners? Which of these we choose is unimportant, but we must find a way to establish ourselves in this space because in the super-disintegration business environment, Taiwan possesses excellent technology, and if it doesn't have a market in which to make effective use of it, the benefits will be compromised. In order to make effective use of it, providing special rights to local companies, cooperating with foreign firms, or even doing it on our own are all viable options. The important thing is that in the process, you must get a grasp of more knowledge and skills, then move forward cautiously.

The mainland Chinese market is not mature, and there are many uncertainties. If you want to enter that market, you must make sure to target areas that will not change. For example, in creating a working team in China, there are many factors outside your control. However, if what you're after is creating technology, or developing software, the risks are much lower—this area is not affected by political changes. You must consider business conditions in China and decide the priority of the areas in which you will invest. Even if you will be entering joint ventures with local partners, you must make sure that you have your own front-line executives, regardless of whether they're people from inside the company that you dispatch or local people that you train. This is because such personnel will be your point of reference for developing relations with business leaders and future talent in China; this is an investment you must make.

Getting B2C customer relations management right

Q: Even now, B2C e-commerce worldwide is still losing money, and that includes even Amazon.com, which has accumulated losses of hundreds of millions of dollars. What is your view of the prospects for B2C?

A: B2C can be divided into two parts. The first is physical items, and the second is intangibles. There's another way to categorize it, which is new things on the one hand and old things on the other. However, there are not many new things because people don't need too many new things in their lives; what's new is mostly just the form of transactions and the delivery of goods.

Therefore, B2C is at heart a matter of asking whether what's being sold is physical objects or intangible data. For example, music, e-books, software, video games, or even fund transfers, are intangibles, which can be handled through mechanisms on the Internet, giving rise to the opportunity to pioneer new markets—this is one type of business model. Another type of model is exemplified by books and ticket sales, where both physical items and information bits are bought and sold; in the future, it will probably be possible to deal with just non-physical data.

If B2C is based on the sale of intangibles, and the e-commerce mechanisms are complete, costs will be reduced to a minimum. If physical items are being sold, it's less possible to lower costs, and the advantage of this business is not derived from the transaction process, but from the total automation of the Internet. This automation makes it easier to optimize operational and inventory controls, which can also help reduce costs. This is the basic reason why Dell's effectiveness has increased as it changed from a direct marketing to an online marketing company with zero inventory—it gets the money first and then ships the product to consumers. From this perspective, the brand recognition, customer information, and transaction mechanisms that Amazon.com's online bookstore has accumulated are all very valuable, and adjusting things to make the company profitable is no problem.

The most important thing for B2C is managing customer relations well. What is your brand positioning? Once customers use your services, what is your relationship with them like? Will they feel satisfied? These issues are the most important. The scope of the Internet is enormous; companies must concentrate their efforts on particular areas and become the best in the world at them. To take advantage of the numerous opportunities presented by the super-disintegrated business environment of the Internet, Acer is in fact developing plans for a virtual company, and many tasks will be performed based on a virtual company model. At present, we are in the process of thinking this out and hope that we will be able to take advantage of some new opportunities to innovate.

Q: **Barnes & Noble bookstore and Amazon.com are quite different. Barnes & Noble established its Web bookstore to supplement its original business model. Which company's approach is more effective?**

A: Based on the Web business model as it now exists, the more pertinent question is whether the value of your original business can be increased by exploiting the Internet. Doing business on the Web means having a vision because without one your road will be rockier. This vision does not cover the next five or ten years, but concerns what basic infrastructure or environment you want to create in two years. You must move forward step by step, and only with a vision can you stay a step ahead of everyone else.

CHAPTER 3

Internet Organizations—Organizations tuned for the knowledge century

THE ERA OF THE DIGITAL ECONOMY is fundamentally based on knowledge, while at the same time it is seeing the introduction of the super-disintegration business model. In response to the arrival of this new era, we must use the internet organization approach, with stress laid on effective resource management.

There are many varieties of network organizations, and the client–server structure that Acer once promulgated is one of them. However, the most advanced, most synergistic, most manageable of all is the internet organization. Though networked organizations have been widely discussed in academic circles, the internet organization is likely to be the most effective, even if it is not fully mature conceptually because many of the management protocols need more time to complete. However, the situation can be likened to a computer needing an operating system in order to function, while real-world applications need the relevant software. Keeping this perspective in mind, the Acer Group has already been implementing the internet organization concept.

A NEW CENTURY, AN INFORMATION CENTURY

The notion of interlinked networks is used in examining organizational structure because the broader business environment has already changed so much as to make old ways of thinking about organizations obsolete. The digital economy is one based on knowledge, and one that uses the super-disintegration business model. As business confronts the challenges of this new era, the internet organization approach, with its stress on effective resource management, should be proposed. In the past, discussion about resources in the traditional economy meant water, precious minerals, land, and other physical resources. As humanity uses these up, these resources are becoming more and more limited. However, there are two types of resources that, with increasing use, actually become more difficult to deplete: one is ever-increasing computing power, and the other is human expertise. Human capabilities and expertise are rapidly growing, and this represents a qualitative change from what was true just a few decades ago.

In the past, many management ideas were based on concepts from the traditional economy, with various theories all having the built-in assumption that resources were limited. Now, with computing power increasing and human expertise more plentiful, our thinking must be adjusted. Computing capabilities and expertise are resources that cannot be depleted. Computing power has developed from early mainframes to PCs, and now to the Internet. Expertise employs education, the spread of democracy, along with the development of information and communications technologies, to give rise to a rapid growth in numbers of talented people, and moreover their presence everywhere in the world.

THE KNOWLEDGE ECONOMY

As the traditional economy is transformed into a knowledge economy, the center is shifting from physical labor to human expertise, from the visible to the intangible, from manufacturing-oriented to marketing-oriented, from hardware to software, from a focus on efficiency to a focus on leadership. Looking at things from this perspective human talent is the most important resource, in the knowledge economy. Moreover, as the nature of the economy changes, successful formulas that worked in the past may no longer be viable, and past experiences may not necessarily provide a valuable reference, but may even turn out to be an impediment.

In the New Economy, intangibles will be more and more important, and companies will rely on innovation to create various types of value. Intangibles will become extremely valuable, and the ability to disseminate these intangibles widely will lead to corresponding profits. This is the reason for the incomprehensibly high stock prices that many Web companies attained in the late 1990s; their nature is fundamentally different from companies that make up the traditional economy.

THE SUPER-DISINTEGRATION CENTURY

In the super-disintegrated knowledge economy, precision management is extremely important. If a business cannot be the leader in a chosen market niche, it might as well relinquish that business. However, there's no need to get too alarmed if this happens; even if a company loses out in one area, there are many others to target and try to take a leading position in.

The emergence of super-disintegration means that the most standard and open protocol or specification becomes an extremely important foundation for business. The main objective of such standards is to allow companies to more easily meet various new market requirements as they develop.

In the recent evolution of human societies and industries, the Industrial Era has developed into the Information Era, which then metamorphosed into the era of the knowledge economy. Industrial development has also evolved from the predominance of vertical integration to disintegration and then to the super-disintegration of today (see Table 3.1).

Table 3.1 The organization evolution

Economy	Industry	Organization
Industrial	Vertical	Integration/hierarchy
Information	Disintegration	Flat/empowerment
Knowledge	Super-disintegration	Network

© 2000 Aspire academy. All rights reserved.

As computing power and talent become ever more accessible, organizations evolve. Organizations can be classified as hierarchical or networked. For a hierarchical organization, the larger the organization, the easier it is for disjunction to occur, where the higher levels don't understand the lower levels. Making decisions may require a very long time, and the problem of high operational expenditures can easily arise. As a result, the organization evolves into a flattened structure with greater reliance on delegation at each level as a way of avoiding the problems that are inherent in a hierarchical organization. In actuality, the concept of a network organization was introduced some time ago, but there is as yet no definitive conclusion about what makes an effective network organization, and this is an area for further investigation.

SPECIAL CHARACTERISTICS OF A KNOWLEDGE ECONOMY

A knowledge economy possesses several distinctive features, the foremost being that the nature of your task is always changing, and that it is multi-faceted. The market is diverse and variable, while time frames rapidly shrink, all leading to additional variables to consider in the management of an enterprise. Next, intangible value becomes more and more important, giving rise to new categories of high value-added businesses and continually challenging traditional management practices.

From Figure 3.1, it can be seen where the problems in a hierarchical organization arise. The graphic on the left side of the figure is the organizational chart familiar to everyone, where the more levels there are, the more complicated the structure becomes. To compare this structure with that of a network organization, each operational unit in the traditional hierarchical organizational structure can be visualized as a circle (see Figure 3.1). From the large headquarters arise smaller local headquarters, and so on, creating a complex and minutely detailed organization (see Figure 3.2). An army, government organization, and manufacturing operation are all examples of this type of organization. For simple and repetitive tasks, this structure is viable; but for highly variable and diverse tasks, it is not effective.

Network organizations (see Figures 3.3 and 3.4) use OP (organizational protocols). And this type of organization looks like a business group, but it is a virtual organization. Running the organization relies on protocols and communication channels between responsible parties. These protocols must be approved by virtual headquarters, and in the new breed of organizations, headquarters are all relatively small. Looking at the situation at present, organizations with large headquarters are ineffectual; regardless of the particular tasks they face, large headquarters are unable to operate effectively. Unfortunately, headquarters, like any bureaucracy, typically grow larger and larger with time, unless they are purposefully restricted from doing so.

Internet Organizations

Figure 3.1 Hierarchical organization and management

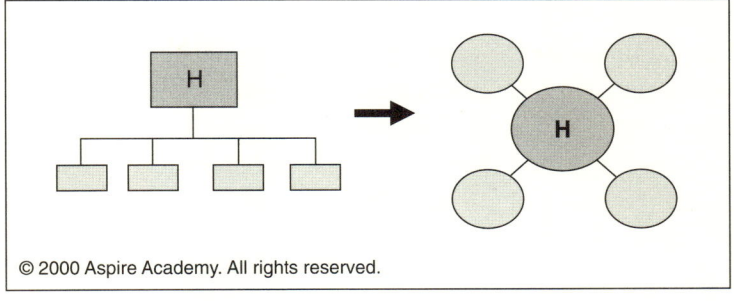

Figure 3.2 Hierarchy in large scale

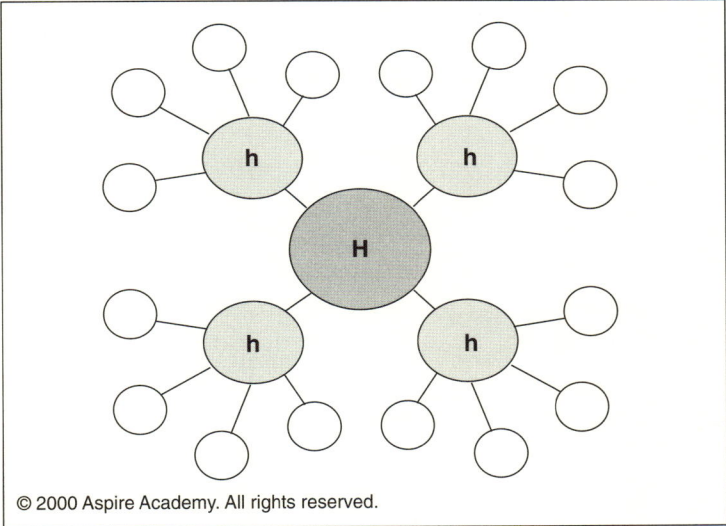

In a network organization, actual operations are carried out by independent units within the organization, which is virtual in the sense that there is no necessary correspondence between its structure and the physical location of individuals or working teams. A large business group has large virtual headquarters to control operations, with subgroups having small virtual headquarters to manage things. Therefore, the actions of individual headquarters are quite limited. Outside each network are different units, and this is done to delineate operational characteristics, with a few special protocols added to help define their interaction. These small networks then combine to form the internet organization.

Looking at Figures 3.2 and 3.4, both hierarchical organizations and internet organizations appear to be quite complicated. What is it that distinguishes the two from each other? Figure 3.5 on page 43 helps clarify their differences.

When Company A and Company B in a business group want to work together on some task, if the organizational structure is hierarchical, then the group's main and smaller headquarters must be involved and give approval. The biggest difficulty lies in the fact that the people who really understand the task and are specialized to undertake it are in Company A and Company B, but when the case reaches group headquarters it may

Figure 3.3 Network organization

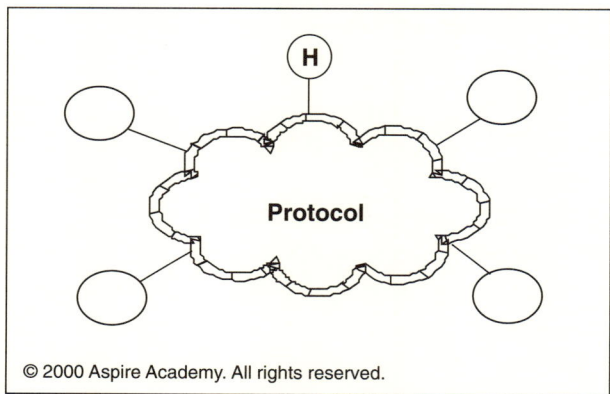

Figure 3.4 Large-scale network organization: the internet approach

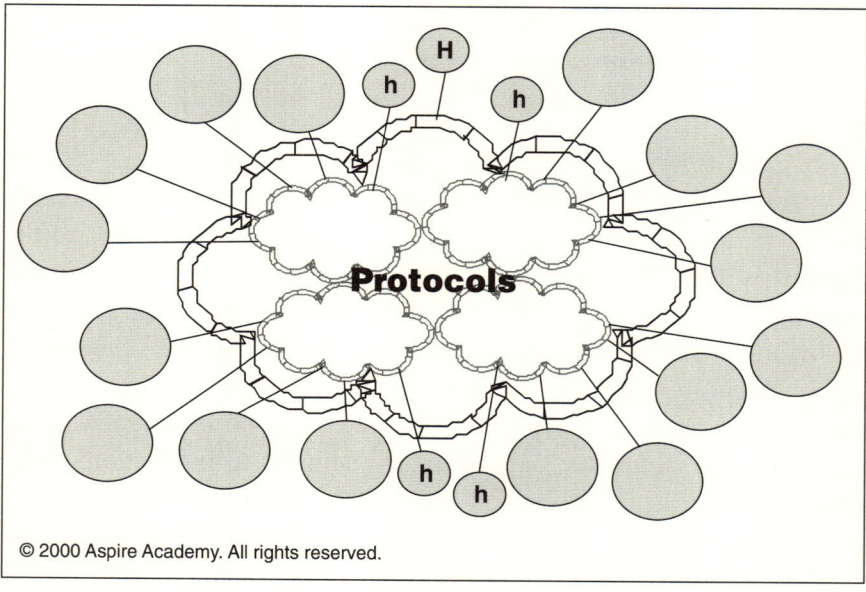

INTERNET ORGANIZATIONS 43

not be the true leader who makes the decision, but his or her advisors. Thus, during the process, people who are not the best qualified end up making the choices.

In contrast, looking at things from the perspective of Acer's internet organization, A and B communicate directly, and the procedures they follow conform to the group's protocols, which include brand usage, corporate culture, information system architectures, business ethics, and so on. Nothing else requires the involvement of headquarters because Company A and Company B are independent entities and as they must each take responsibility for their decisions, they should communicate directly.

Moreover, in a knowledge economy, tasks are more diverse and changeable, time frames are short, and every day in the operation of a business is different. A hierarchical organization cannot react fast enough, while an internet organization is much more nimble.

In order to meet the needs of the market, assume that Company C wants to undertake some change or staff adjustment. In a hierarchical organization, the participation and approval of headquarters is required; however, in an internet organization, C can make these decisions on its own (see Figure 3.6). The reason is very simple—each company in the group is independent, with its CEO deciding on policy. At most, the board may make a resolution; or if the board doesn't perform well or

Figure 3.5 Hierarchical vs internet organization (1)

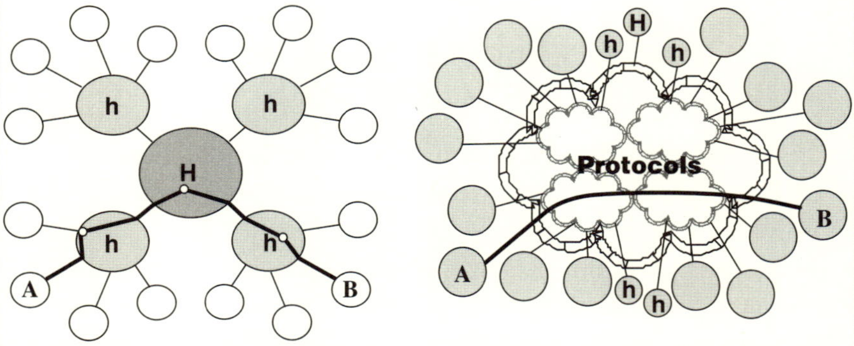

A and B's cooperation must be processed through H and two h's cooperation

A and B's cooperation does not need to pass through H or h

© 2000 Aspire Academy. All rights reserved.

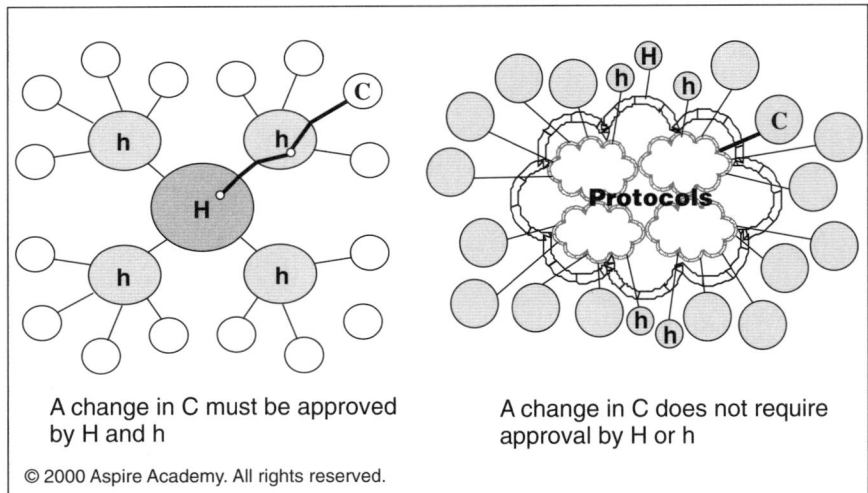

Figure 3.6 Hierarchical vs internet organization (2)

cannot come to a consensus, the issue can be resolved at a shareholders' meeting. This entire procedure takes place independent of headquarters and can't be controlled by the parent company.

Acer's way is, as far as possible, to not allow the parent company to control more than 50% of the stock of any member company. This is very different from the thinking in most organizations, where the majority of parent companies want more than half of the stock in order to maintain control. The attainment of control in a network organization does not depend on the size of your stake, but on shared interests and organizational protocols.

CHARACTERISTICS OF AN INTERNET ORGANIZATION

In an internet organization, each unit is independent and specialized in its particular area of business. Headquarters does not necessarily control all businesses. Within Acer, long ago I began enjoying the pleasures of relinquishing power, but many people don't understand the situation; if they want to make a purchase or recommend someone to join the Acer Group, coming to me is wasted effort. That's because headquarters doesn't have that power; each organization within Acer is responsible for itself, and this has been the case for many years.

As an organization gets larger and larger, the working units within it become more and more numerous. For effective management, neither a flattened organizational structure nor a system of delegation can match an internet organization. When a network organization grows to a point where it is very large, it must give each small network its own protocols to control interaction since it is easier to manage small networks. These small networks are then linked together in a large network, and that is what an internet organization is.

Each network can manage itself within the scope of its operations and perform its functions and services well. However, if many networks are to be interconnected, their interaction must be managed through effective protocols to allow them to integrate smoothly.

What exactly is the key factor in a successful transition from a client–server organization to an internet organization? Besides a network organization's protocols, the "inter" part of the formulation is vital. In the real world, using a single monolithic network to link everything is not possible, unless each individual region has its own robust network.

Advantages of an internet organization

In knowledge-based industries, companies must be both diverse and focused. If they are not focused, they will lose out; however, if as soon as they become focused, the market changes and business opportunities change, they have to find a way to appropriately diversify and have a diverse range of competencies, otherwise the room for survival will be squeezed. Therefore, companies must use the virtual dream team model and allow working groups to be flexibly composed. The fastest method is the internet organization, where everyone follows standard protocols and can rapidly come together and arrive at a shared understanding. The people on the dream team can develop the rapport to face various challenges effectively, and the internet organization thereby achieves a balance between diversity and focus.

In addition, internet organizations are more effective thanks to their speed in decision-making, effective composition of working teams, and ability to adapt to changing conditions. Such organizations possess flexibility and are skilled at knowing at what point to stop; when they reach a certain scale, they start diversification and spin the diversified part off as a new node in the network, the equivalent of a website on the Internet. The process takes place because if a single website handles everything on its own, it cannot sustain sufficient focus, which is essential. A successful website makes everyone in the whole world think of it as the authority in the particular area that it's focused on, and the same principle applies to companies operating in the super-disintegrated business environment.

To meet the needs of a changing, diversifying market, different types of working teams must be recruited and the organization continually re-engineered. In addition to being the most effective type of organization for performing these tasks, the internet organization can lower the operational costs of headquarters. The operational costs of running headquarters can be enormous. There was a period when Sony's headquarters experienced a rapid growth in staff strength, causing its competitiveness to decline, which led to the decision to reduce the number of people at its headquarters. In 1990, annual revenues at Acer were less than US$1 billion, but there were more than 300 staff at headquarters. Now revenues have increased tenfold, but there are only 100 people at headquarters.

Even more importantly, an internet organization is the most natural form of organization. Human societies are themselves like internet organizations. In English, *legal entity* and *individual* have distinct meanings, but in Chinese the equivalent terms both refer to people, implying that both retain their right to be treated as human beings. Legal entities cannot last forever, but their spirit and traditions can be handed down through internet organizations. In their daily lives, people operate quite autonomously, and when they do interact with others it is in a very flexible manner—playing different roles in different groups. Internet organizations more closely reflect people's natural social interaction than do strongly hierarchical organizations. An organization may not always have the same objectives and working teams, but through dissemination of fundamental ideas and adaptations to changing times, the organization can gain new life, new missions, and the ability to replicate itself.

CHALLENGES FOR AN INTERNET ORGANIZATION

You have no other choice—you have to use one type of organization or a combination. An internet organization naturally faces many challenges as well. First, protocols must be clearly defined. It's not difficult to discover that the situation is similar to that in trying to nurture the ability of the members of a sports team to play well together: many protocols are intangible, impossible to describe. A brand is an intangible resource, and in each company within a business group the brand is relied on in getting certain tasks done, while at the same time all efforts are made to perform the task well and enhance the brand image. For example, say that a company's image is one of transparency, integrity, and protection of the small shareholders' interests. These things become part of a tacit understanding that the members of the business group have a type of organizational culture; by having a common culture, everyone can do their job more easily.

An internet organization in a knowledge-based economy naturally cannot escape the need to establish an effective information system infrastructure. The tasks that can be handled by such a computer infrastructure are numerous, but the fact remains that computers are just *things*. All computers must comply with network protocols; but it's difficult for people to always conform to set rules, and sometimes even the best of people will violate tacit understandings in performing a task. The operations of a business are much more complicated than a computer processing data, which is why computers are found lacking when they are asked to handle human matters. In fact, this is not a problem of the computers—the processing tasks that an organization handles with computer assistance are themselves ill-defined. The protocols must be clearly defined, and this is best done by a committee formed of people whom everyone in the organization supports.

The most problematic feature of a network is that control is not localized at a single point, and this fact is reflected, for example, in the many undesirable developments on the Internet. There are two schools of thought contesting each other on the issue of how to deal with these undesirable elements: one advocating pre-emptive control and the other arguing against interference. Negative developments will also occur in organizations, and when they appear, should efforts be made to control them, or should they be allowed to run their natural course?

When dealing with a business or task, integrating so much talent and resources requires an organization, whether it is a hierarchical, client–server, or internet organizational structure, or sometimes a mix of different types of organization. For example, the Acer Group is an internet organization, but the group's companies, despite being flat and relying heavily on delegation, may still be hierarchical organizations. Manufacturing sites are multi-leveled hierarchies, while sales operations have fewer levels; so for different tasks, different organizational forms may be combined.

INTERNET ORGANIZATIONS IN THE KNOWLEDGE ECONOMY

The internet organization is an effective approach for companies in meeting the challenges of a knowledge economy and the Internet age. This area is exciting because I am personally involved in using internet organization principles in running the Acer Group and integrating the distinctive features of highly entrepreneurial Asian companies. If people research the world's most advanced new organizational modes, Taiwan has found a place with its formula of "using small size to win big."

The internet organization is especially suited to companies wanting to be niche leaders. For an organization to be effective as a niche leader, it requires diversity in addition to focus; this is because the market that a niche leader targets is likely to be too small. The internet organization can help in balancing diversity and focus.

A company can use an internet organization structure to build up the larger scope needed to compete internationally. In the Internet age, even business groups from different industries may combine to form a larger new group, and this may be a new competitive mode that is a perfect match for an internet organization. In this arrangement, each company within a group would be a largely independent entity that takes responsibility for itself in facing the needs of the disintegrated global business environment.

Typical Japanese *keiretsu*, Korean *chaebol*, and Taiwanese conglomerates all use hierarchical organizations, and they will find it very difficult to adapt to the diverse and changeable markets of the future. Even at present, the return on investment for Japanese conglomerates is not very high and cannot create much benefit for workers—a weakness that internet organizations avoid. The Korean *chaebol*s are large, but strikes are frequent because they cannot satisfy the desire of workers to prove themselves; internet organizations give workers with ability the opportunity to demonstrate their potential. They do this by giving people more autonomy and allowing them to play more flexible roles, while also encouraging entrepreneurial efforts within the organization.

It is worthwhile to apply my competitiveness formula (see Chapter 5) to these issues. If greater competitiveness is not achieved by cutting costs, it is done by creating value. There's more room for cutting costs in a hierarchical organization, while internet organizations are better suited for creating value. In the future, economies will be knowledge-based, meaning that there will be opportunities to create value, but, more importantly, after the creation of value and if it is done correctly, costs will not increase while value will continue to rise and competitiveness will thus be greatly enhanced.

In the final analysis, should companies try to sustain the old economic models or welcome the New Economy? Many companies can easily adopt the internet organization model. It's not only easier for many companies to take this route, but taking the hierarchical organization approach would be very ineffective. In the future knowledge economy, a company of maybe 50 or 100 people may control the world's largest market in its area of focus—and it won't be only large manufacturing companies with thousands of people that are competitive.

Conclusions

Internet organizations meet the needs of a knowledge-based economy and super-disintegrated business environment. The internet organization is an extremely unique, effective model that can aid many businesses in raising their international competitiveness. Many people complain that Taiwanese companies' willingness to be niche leaders is a negative attribute, and many people have hoped that the government would follow the lead of the South Korean government and actively cultivate a large world-class enterprise. However, that is an old mode, and our ethnic characteristics make it impossible for us to do that. The internet organization concept can allow Taiwanese and Asian businesses in general to compete effectively internationally.

Just as a society must be dynamic and sustain economic progress, democracy and the rule of law are necessary preconditions for effective internet organizations to emerge. An internet organization is a democratic organization ruled by law, with democracy meaning in this context that each company maintains the equivalent of human rights, and the rule of law refers to shared protocols.

In an internet organization, each company is like a website, being independent and responsible for itself, while focusing on the most optimal contribution it can make in its chosen field. Because each company is independent, but has the ability to work well together and conform to protocols, the organization as a whole can develop rapidly. In an internet organization, shared benefits for all parts of the organization must be established in order to make effective management possible.

Organizational leaders must subscribe to the principle of power-sharing; not only this, but they must be willing to relinquish power. The situation can be likened to a family. If the family is to flourish, the head of the family must be willing to let the children be completely independent, allowing them to fully develop their potential. The Chinese folk wisdom that family wealth can only be sustained for three generations assumes a hierarchical organization for the family. If an organization is not hierarchical, it should be possible to sustain its prosperity indefinitely.

The origins of the Internet go back to the 1960s, and it has only begun to become pervasive in the last five years. For something to go from an initial appearance to exerting a broad influence requires a very long time. The internet organization concept is still very rough, but such an organization will not only offer lower operational costs than traditional organizational structures, but it will also prove to be more effective as a whole in the knowledge economy.

Discussion

Network protocols: seeking positives rather than avoiding negatives

Q: **The key factor in building a successful internet organization is probably higher management. If executives follow protocols superficially, and in fact violate them, how can this be addressed?**

A: There are two levels. The first is the level of culture. Taking Acer as an example, its distinguishing feature has been the creation of the big from the small, and Acer's corporate culture has always been oriented in this direction. If an executive does not comply with the directions set by the corporate culture, not only will people higher up apply pressure, but even staff below the executive will give him or her pressure. The natural result will be a sense of oppression, which will make it impossible to violate the tacit understanding.

However, tacit understandings are not set in stone, but just offer a general direction—they don't aim to specify that these particular actions are right and those are wrong. As has been emphasized repeatedly in the past, the principles underlying a corporate culture are unchanging, but the way they are interpreted can be constantly adjusted. If an executive obviously violates a protocol—for example, violating the rules of a brand identification system—someone in headquarters who is devoted to brand management will directly inform the person involved that corrective measures must be taken.

When something becomes an obstacle for other people, it's like the appearance of a computer virus; the virus has to be eliminated, and based on the characteristics of the particular virus, improved anti-virus technology must be developed. Protocols are like this—after problems constantly emerge in an organization, a committee develops a new, more effective protocol. A key point is whether the setting of protocols aims to seek out positives or to avoid negatives. Network protocols should take the former as their starting point—just as we cannot stop pornography on the Internet, we cannot on this account give up using the Internet because the positive advances it makes possible are just too great by comparison.

Corporate culture can serve to entrench internet organization protocols

Q: **Internet organization protocols rely on tacit understanding of their underlying principles, but after a large increase in the number of people at a company, how can this understanding be cultivated? Is it culture or shared interests that must be relied on?**
A: The problem lies in whether there is room for choice. For an organization to sustain development, there are two choices. The first is to use a hierarchical organizational structure and the other is to use a network organization structure. If both types of organization will give rise to problems, then the question is which type will be more effective at resolving the problems. For a culture to encompass common protocols for everyone in a business group takes time; the level at which a single person can exert influence is extremely limited, but in propagating a corporate culture, an internet organization structure is more effective than one that is hierarchical.

However, corporate culture is a very broad concept, and even though principles can be set, they will have different interpretations because each part of the organization is dedicated to different tasks, at different operational levels. The internet organization concept can first be used with easily managed individual organizations, and then protocols can be applied to the relationship between these smaller internet organizations, to attain effective operation of the whole.

From the point of view of the Acer Group, each company within it can operate well on its own, and this is far better than a single large but ineffectual organization. It's as if you are asked which of two businesses you want to choose. Do you want the group made up of 100 companies, with 80 doing well and 20 running less than ideally, but all striving for the glory of the group as a whole? Or do you want the monolithic conglomerate, big but unable to effectively manage itself? Faced with this dilemma, I chose to establish the internet organization mode.

We cannot guarantee that no one will break rules. Acer's corporate culture is not necessarily followed by every executive, but it's probably the case that people lacking in confidence cannot adhere to our "share know-how completely" corporate culture. All that can be said is that in an internet organization, it's less likely that talented people will get lost in the shuffle. Acer Group headquarters has dealt with companies in the group that broke rules—the corporate identity system (CIS) is an example that is visible, and there are many that may not be so visible, such as business ethics. Once, an Acer Group company was very eager to go public, but we felt the time was not yet ripe, so we stopped it. This action was taken based on a very important protocol because it would have had an impact on the group's image.

Managing resource access, rights, and responsibilities

Q: **In a network organization, can a particular part utilize a hierarchical structure? How can resources, rights, and responsibilities be clearly defined?**

A: This has to be divided into two levels. Below the level of the overall company organization, a hierarchical structure applies—though as much as possible the organization should be flat and highly delegated—with the operational units being treated like network nodes. A distinctive feature of internet organizations is that each node is fully functional. In the past, Acer was organized into strategic business units (SBU) and regional business units (RBU), and used a client–server organizational model, but the internet organization model could not be applied to them. This is because SBUs and RBUs were not fully autonomous and could not be managed on a client-to-client basis; as a result, they had to be combined into global business units (GBU).

As for defining resource access, rights, and responsibilities, a system of "brothers each paying their own way" is used. Each company is an independent entity, and material resources are handled through normal business transactions. Ownership of intangible resources may be ambiguous, so they are shared as much as possible. Taking promotion of a brand as an example, each company contributes a portion of its profits to headquarters, headquarters then allocates funds for promotional activities, and the results must then be distributed back to each company.

We specify that each company in the group must devote a portion of its budget to activities to publicize Acer and enhance its image; everyone shares in this responsibility. Intangible items must be shared, which entails setting protocols—in fact, this is the most difficult part. However, in a knowledge economy, these intangibles are the most crucial—whether management knowledge, culture, or image—they are the key factors in a business's development.

In establishing Acer Academy, Acer's headquarters required every Acer company to contribute funds, the academy in turn provides opportunities to receive education. However, there was a degree of freedom; if you paid, you didn't necessarily have to go to class, even though some people expressed appropriate concern about this. We have many regulations that aim to capture a spirit and preserve flexibility. As for rights and responsibilities, each company manages itself, with committees having the obligation to set some shared protocols.

Effectively resolving the problems of multiple units

Q: **In such tumultuous times, can the internet organization concept really handle all the details of running an organization?**

A: The business environment is very complicated, and the question is what method you use to solve problems. You can expend much energy or use complicated structures to take the right actions in dealing with complex problems, but once you've done it, perhaps no one will understand it. Moreover, the complexity we're talking about is multi-faceted and constantly changing. The idea behind internet organizations is to establish common interests and shared protocols, and to try to make it very easy for everyone to do their job when they face complex situations. I have also spoken of the fact that the ability of every person to handle various types of situations, namely, expertise, has already been significantly increased. Internet organizations suit the needs of knowledge-oriented businesses, as well as meeting the demands of the super-disintegrated business environment.

Q: If there are many companies within an internet organization, how can they be effectively managed?

A: The internet organization concept specifically addresses the problems faced by an organization made up of many complicated parts. In a hierarchical organization, if there are many parts the organization becomes made up of a large number of levels, causing problems to develop in communication between higher and lower levels. When a problem arises in the lower levels, the upper levels may not know about it, and thus will not be able to make adjustments in a timely manner. An internet organization can effectively resolve this problem; it can take several units that need to be managed and turn them into a network. Then some protocols are set, with the condition that these protocols do not violate the protocols of the larger group. In this way, the problems arising from proliferating parts in an internet organization can be resolved.

In the future, it is highly likely business will more and more require that many different operational units be coordinated. That's because compared to the past, markets in the future will be more diverse and changeable, and this will be true for every industry. Therefore, traditional hierarchical organizations will not provide what's required and will only be able to effectively handle a small range of simple, repetitive tasks. In the past when Acer employed the client–server business model, even though each operational unit was independent, none of these were permanently sustainable. Internet organizations give rise to independent, permanently sustainable operational units, and the advantage is that these units do not have a dependency mindset; they must compete and constantly improve and develop. A crisis mentality keeps the organization from stagnating.

Q: The creation of protocols is done by committees, but is it not possible that the members of the committee could set inappropriate protocols?

A: The main objective of a committee is to find the common ground in a large business group and then create shared protocols based on this common ground. It must be stressed that committee members must be representative of the group, and there are two types of representation. The group as a whole may be represented by the highest-level CEOs in subgroups. In these subgroups of CEOs—for example, those who do Web services—many will be young people, and they will be able to devise a small-scale protocol that applies to their particular operations. However, they must ensure that these protocols don't conflict with the larger group's protocols. Through the committee process, which can accommodate specific needs of individual companies in the group, as well as the general interests of the group as a whole, they can create protocols that suit the tasks that the organization faces.

The Cisco mode versus the Acer mode

Q: **The internet organization concept can be understood by comparing two companies. Cisco uses the acquisition of smaller companies to create dream teams, while Acer uses the creation of small startups to build up an internet organization. Which mode is more effective?**

A: Cisco's and Acer's cultures are different in some ways and similar in others. Cisco applies centralized management while Acer utilizes distributed management, but they both underline the importance of knowledge and resource-sharing. Cisco is aware that a single company is unlikely to develop all the products and technologies necessary to satisfy its customers in the Internet era; therefore, Cisco continually acquires potential startups to maintain its large market share in the U.S. and the world. Cisco is a large organization with a synchronized corporate culture and information system. When Cisco acquires a new company, it deploys people from the parent company to help personnel quickly adjust to the parent company's culture. Cisco also requires the information systems of the acquired company to conform to Cisco's, so the acquired company can enjoy the benefits of Cisco's digital infrastructure. These approaches have proved to be an effective way of keeping Cisco's efficiency and unity after a number of acquisitions. Acer's model is not only about creating startups, but allowing these startups to operate independently. These new companies share our pedigree, but our hope is that they will grow bigger and bigger, and even more that they will do their job better than mother Acer—we push them to move outward. We also have other groups outside that produce new Acer companies, through venture capital investments or strategic alliances, to extend the Acer internet organization.

Cisco's approach may be more suitable for the U.S. market, which requires intense and consistent focus. However, when a company does not have a large monolithic home market, the internet organization approach provides greater flexibility for meeting the needs of a large number of smaller markets.

Protocols allow the group to operate effectively

Q: How can group company CEOs utilize shared protocols to create shared benefits? If an individual operation violates a shared protocol, what mechanism can be used to correct this behavior?

A: Shared interests can be divided into tangible and intangible types. The executive teams of all the operational units are the company's chief owners, and the largest actual interests belong to the people who run the company; they have a motivation to help the company perform to its fullest potential. The CEOs of subgroup companies will have a stake in the performance of other companies, and this refers not only to operations, but also works through the possession of stock; so they will also hope to see other companies in the subgroup perform very well. The performance of the group as a whole is related to the personal stake that high-level executives have, and this relationship is based on actual shared investment.

An intangible benefit is the shared interest provided by the group brand. Having a brand is inevitably more convenient, in that it creates an advantage in everything from relationships with banks to sales and staff recruitment. Each company must conform to the image of the brand definition in its dealings with the outside world, as well as work to strengthen the brand, with the idea being not just to passively enjoy its benefits, but also to give something back and contribute something toward enhancing the brand. In this way, protocols arise naturally from the process of working out how best to serve the interests of a stronger brand.

Q: **Internet organizations use the Internet for communications, but does this dilute the group's own culture?**

A: An internet organization uses Internet technologies to allow knowledge-sharing and to implement collaboration tools. An internet organization is developed slowly—it's not a matter of forcibly taking twenty companies and instantly melding them into an internet organization. There must first be a core that gradually develops outward, propagating its original business culture and using internet organization protocols, which makes resource-sharing more effective. We have some companies outside the group that have developed on their own, with less of the original company integrated; however, the performance of these companies is less ideal, and from this we can see that the protocols are an important factor in the effective running of a business group.

Q: **In the Internet economy, there are large and small companies, and whoever possesses the superior culture will be able to attract more talent and create greater gains. Is this description accurate?**

A: The most important point is not who wins and who loses, but the progress of society as a whole. In the past, knowledge could not earn its possessor much money, but in the future, knowledge will be economic power; a knowledge-based economy can be used to enrich a country, strengthen a country. Perhaps even smaller countries in Asia will be able to call their own shots in the future.

CHAPTER 4

A New Vision— High-tech with a human touch

TAIWAN: A MODEL FOR ASIA

NEWLY DEVELOPED ECONOMIES should seek excellence in both hardware and software, and this fits in with many concepts that have been discussed in the past as directions for Asia, as "culture and technology island" has been for Taiwan, where culture is software and technology is hardware. Software and hardware must be balanced and developed in parallel; after countries have developed technologically, they must stress a balance between culture and technology. In addition to Taiwan and several other countries in Asia facing this transition now, many other countries, a little behind in terms of economic development, will have to deal with this issue in the near future as well. Taiwan's experience can thus serve as an important reference.

WHY A NEW VISION IS NEEDED

The post-PC era is already approaching, and with the development that the Internet has undergone over the past few years, a new digital economy led by the U.S. is taking shape all over the world. Returning to my smiling curve concept (see pages 10–11), you can see that the left and right sides are where the greater value resides. At present, Taiwan's information industry, and most of the rest of Asia's, is situated very firmly in the center. At this critical juncture, we must use the base we have established and move as quickly as possible toward the left and right sides, strengthening intellectual property assets, software, and services to create value. This is the most urgent matter at hand.

Asian companies have proven themselves adept at hardware manufacturing, but, with a few exceptions, lag behind noticeably in software. There are many reasons for the shortfall in software, such as local markets being too small, but there will be opportunities in the future. What we should guard against is not being aggressive enough about developing application software; if we just stick with the status quo, in the future there may not be much room to add value. The relatively poor performance of Asian companies in software is not so much a matter of inadequate capabilities as a disadvantageous mindset.

In the U.S., software does not play second fiddle to hardware; it leads the industry aggressively. By contrast, in most of Asia, hardware definitely plays a dominant role. The irony is that as a whole, software workers are more skilled than hardware workers. This problematic attitude must be corrected—hardware and software sectors should effectively complement each other and cannot simply be allowed to develop independently.

Furthermore, much of Asia's software industry is the result of investment by foreign companies, with a view to supplying local and not international markets. Worst of all, the software industry is largely focused on winning government contracts; and government contracts are the "least efficient" business because fulfilling one contract takes many years, and on completion the know-how gained cannot be recycled. Fortunately, the situation has begun to change. Because usage of the Internet in business and by private citizens has become so prevalent, the market has enlarged, and the profit potential and development prospects for software companies in Asia are looking much brighter.

BALANCED STRENGTH IN BOTH SOFTWARE AND HARDWARE

The Acer Group held a strategy meeting in 1997 that set forth a vision of "creating human-touch bits" as an answer to the state of software development we saw for the year 2010. The objective was for one-third of the group's profits, and one-sixth of its revenues, to derive from its software business. The goal set at the strategy meeting might have seemed very general, but it is very simple and, after some necessary refinements, it left a few key points of consensus that established a consistent direction for everyone to work toward.

Recently, the chairman of the Taipei Computer Association (Taiwan's most influential private-sector information industry association) and myself, as well as several delegates, held a strategy meeting that came to several conclusions worthy of attention not only in Taiwan, but in other countries at a similar stage of development.

During the strategy meeting, we defined the Taiwan of the future as being equally strong in both software and hardware. This definition fit in well with many concepts that had been discussed in the past, such as "culture and technology island," with culture as software and technology as hardware; "green silicon island," where green represents software and silicon island stands for hardware; and the human touch of software; and the high-tech of hardware. Software and hardware must be balanced and developed in parallel; after a country has developed technologically, it must stress a balance between culture and technology.

There is another point that is very important—in the past, environmental protection was sacrificed for economic development; without this sacrifice, economic development could not have taken place. However, in the knowledge economy of the future, computing power, and the wide availability of talent, should make it unnecessary to sacrifice the environment in order to achieve economic development. The environment is a limited resource, so it cannot be used without restraint. In the new century, damaging the environment will not be a precondition for continued economic growth. This is an opportunity presented by the New Economy.

For Asia to become a world-class innovator in information applications, it will need to use its established information industry as a base, focus on creating new applications, and make development of a world-class knowledge industry its mission. It must innovate in the area of applications. Technology is global; if a new technology is created in the U.S., we don't need to expend effort in innovating in the basic technology itself. The possibilities for applications are unlimited, and Asia has great potential in developing innovative applications.

To become a leader in the global digital economy, information technology must be used to enrich digital content, taking the fruits of human civilization and putting them in digital form, and then encouraging a corresponding economic transformation. The New Economy is more efficient than the old economy, and in the process of renewal, we hope that Asia can serve as a model for the rest of the world.

Strategy for realizing the vision

In order to realize the visions outlined, a few applicable strategies need to be presented. First, the resources of government and business must be adjusted as the larger environment changes, directing them to key areas as they emerge. Second, domestic demand must be created. Economic development in the U.S. is driven by stimulating domestic consumer demand, while most of Asia takes the opposite tack, stressing export sales. To develop the knowledge industry in the future, there must be a vigorous local market in order to fuel growth. Third, business must transform itself, and this includes equipment, software, and staff, with the use of incentives to encourage business to speed up digitalization.

As has been said before, what Asia needs most is international marketing capabilities, and these must be built up through the training of workers with the right skills. In addition to offering courses, there must be other training programs that cultivate marketing skills. Furthermore, a more robust mechanism for interaction between governments and business must be developed. The cronyism that has developed between many Asian companies and their government originated from individual businesspeople meeting with the government, which is not a good mode of interaction. A more prudent approach is for industry leaders to discuss matters and together develop recommendations to provide to the government. Governments around Asia should establish communication channels with businesses that are transparent and that are directed toward achieving benefits for all.

Key factors for success

First, Asian countries must stimulate vigorous domestic demand, speed up corporate digitalization over the next few years, and offer incentives for large-scale investment. Using matching funds to encourage large-scale investments may not require very large sums, but its effectiveness in pushing the digitalization of businesses cannot be discounted. Additionally, traditional industries should be encouraged to quicken their pace of modernization and raise competitiveness. Governments should come up with concrete measures to assist traditional industries in this transition, and one of these measures must include computerization because computerization is undoubtedly one of the key tools for raising the competitiveness of traditional industries.

Second, substantial development budgets for educational and cultural digital content should be created, with private-sector investment leading development directions. With the guidance of the government, the private sector can carry out development work. In this manner, application software and content industries will naturally flourish.

Third, there is already much talk of B2B e-commerce, and the next step should be to provide the environment for B2C e-commerce transactions. There are already some states in the U.S. (such as Virginia) that have proposed tax exemptions for all e-commerce transactions, but this is not a desirable policy. Even though tax exemptions would be a very effective incentive method, they would lead to unfair distributions of tax burdens. Tactics should be based on shifting the target of incentives from the supply side to the demand side, which is very different from traditional investment incentives, where the supply side is the focal point.

ESTABLISHING A DEVELOPMENT STRATEGY FOR A KNOWLEDGE ECONOMY

Coming up with a development strategy for a knowledge-based economy is a matter of great urgency. Though the development of a knowledge-based economy is closely related to the activities of the information technology industry, it is not the only factor. The stage for high-tech development in the future is virtually set, but how to apply technology and develop a new knowledge economy is something that requires the attention of people in every profession and industry.

The road to a knowledge economy has already reached a point where people from non-technical professions and industries are playing the starring roles. It will be interesting to see how they use technology in their creations, and this know-how is one of the keys in the knowledge economy. Since it is the development strategy for the country's own knowledge economy that is at stake, the people who need to be sought out for participation ought to be government officials and policymakers because they control the resources. They should brainstorm visions and strategies, with advisors and executive teams creating specific plans, making adjustments, and implementing measures step by step.

CONCLUSIONS

In the digital economy of the future, the application of information will be a new basis for core competitiveness. A knowledge economy includes many industries, and all must move in knowledge-intensive, high value-added directions as they develop, giving rise to a new kind of economy. In addition, we must work aggressively not only to remove the "wealth gaps" in the digital economy, but also to close the gap between the information-rich and information-deprived. Taiwan and the rest of Asia definitely can take advantage of their capabilities to play a strong role in eliminating the worldwide problem of discrepancies in access to digital technology.

Discussion

Promote innovation through domestic market demand

Q: How should the role of government financial consortiums be adjusted to meet the needs of the software industry?

A: In the area of software, truly global companies, such as Microsoft, were built up gradually, piece by piece. The recommendation of the Taipei Computer Association that innovation should be driven by domestic market demand is a good way of doing things. If business invests one dollar and the government matches it with another dollar to grow the market, then the software industry can develop.

Everyone can enjoy the latest technology

Q: The CEOs of many technology companies have simultaneously begun to express concern about cultural issues. How can technology be enhanced while also eliminating the digital wealth gap?

A: This returns to the issue of basic business philosophy. Why does a company exist? Is it for the sake of earning money, or to meet society's needs? Is the spread of technology a reflection of society's needs? It is very fortunate that this industry has been dedicated to propagating technology and has enabled everyone to enjoy the latest technology in the hope of fulfilling society's needs.

From another perspective, technology coverage in the media also has the effect of stimulating the adoption of technology. The business world also has its obligations; the "idiot-proof computer" that I have been talking about for more than a decade has yet to make its appearance. This is undoubtedly the responsibility of business. If a computer is not backed up with the right services and software, it's not easy to use and has nothing of use; this is the main reason why I stress the shift from high-tech to low-tech to human touch.

Innovation originates from the environment, training, and talent

Q: In order for a country to become a center for well-known brands in the Asia-Pacific region, it must enhance innovation and the aesthetic appeal of content. How do you think this should be done?

A: Is creativity an inborn talent, or something that can be cultivated through training? I believe that the environment is the most important, training second, and talent third. Therefore, the key is people with the ability to take advantage of creative environments. Creative environments can be large or small, and even a home is one type of creative environment. In the knowledge-based industries of the future, the most important thing will be innovation.

Q: How can technological development and environmental concerns be balanced?

A: In terms of business development, environmental costs are getting higher and higher. We cannot destroy the environment in order to develop the economy; this is a principle that must become part of the mindset of business. Through technology and the development of the knowledge-based economy, the old notion that economic development and environmental protection cannot go hand in hand is gradually being overcome. Businesses with the ability must devote more effort toward environmental protection because the impetus for economic development is the improvement of living conditions. Without a good environment, what good is economic prosperity? Now, everyone assents to this idea, but the problem is coming up with a concrete and workable approach.

PART 2

MARKET SHARE FOR CREATED VALUE

CHAPTER 5
A New Basis for Competitiveness—Innovalue

FOR A COMPANY to sustain operations and constantly enhance its competitiveness, the most important factor in the competitiveness formula is innovalue. This is because value for customers is constantly shifting; they love the new and hate the old, and the value they attach to a product changes with time. From the perspective of management and business development, investment in stimulating innovation has the highest return in many different areas.

There are two types of value: first, there is functional value, such as ease of use; and second, there is economic value, which means getting much for a low price. This formula can be used to express simply the level of competitiveness that a company has achieved—the more the value created and the lower the costs, the higher the competitiveness (see Figure 5.1).

Figure 5.1 Stan Shih's formula for evaluating competitiveness

$$\text{Competitiveness} = f\left(\frac{\text{Value}}{\text{Cost}}\right)$$

- Value: service, innovation, quality, image
- Cost: labor, material, natural resources
- Overall perspective: including intangible, indirect, future items

© 2000 Aspire Academy. All rights reserved.

In terms of creating value, from the perspective of the consumer, what this generally means is functionality. However, value can also originate from various services, a reputation for innovation, product quality, and other sources. Usually, when competitiveness is assessed, these items are neglected, meaning that only the visible, the current, and the direct factors are considered. But if the perspective is expanded a little, to include thinking about the future, the intangible, the indirect factors, as embodied in the simple formula, it can provide a very good reference for many tasks and even the setting of company strategy.

Companies around Asia should focus more of their attention on pursuing value and not reducing inherent costs because increases in labor costs and new expenses such as environmental protection costs are just the result of an economy being successful. The objective of economic development is to improve the quality of life, and improving quality means that there will be a corresponding increase in certain costs. The main reason that U.S. economy has performed better than Japan's over the last ten years is that it has played to the hilt the role of creating value. In the past, Japan used automation and mass production to lower costs, but this was not a permanent solution.

DISTINCTIVE FEATURES OF ASIAN COMPETITIVENESS

In the past, Asia's Little Dragons, including Taiwan, used inexpensive labor, raw effort, and industrious workers to create economic prosperity, but these strengths no longer exist or have markedly diminished. However, the trend toward disintegration in industry and the needs of the future all suit many Asian companies' characteristic speed, flexibility, and low-cost brainpower. Speed saves time and therefore money, as does low-cost expertise. And flexibility is the key tool for taking advantage of opportunities.

With all the products on the market, every day sees the appearance of a different competitor, making product quality a must. However, the first thing that consumers notice is innovation: when a consumer feels a new product suits his or her needs and likes it. After that come service and product quality—sometimes these are not related to the nature of the product itself, making aggressive innovation extremely important.

Because too little was written about this issue in the past, too little effort was expended by Asian companies, with the result being a dearth of experience and lack of confidence in the ability to innovate. In the past, competitiveness depended on lowering costs, and lowering costs is still essential for competitiveness in the PC and semiconductor industries. However, Asian companies in the information technology industry now have the ability to contain costs while increasing value, unlike most U.S. companies, where creating value also raises costs. This discrepancy arises because the value created by U.S. high-tech companies is generally derived from breakthrough technologies, which require heavy investment, thus compromising the increase in competitiveness. In addition, because the home market, which serves as the test case for new innovations, is so large, it requires higher overall cost to introduce them. If Asian companies can maintain their ability to contain costs while dedicating their efforts to create new value, there is no limit to their scope for increasing competitiveness.

Six forms of innovation

When speaking about innovation, discussion should not be limited to technological innovation because innovation itself is not restricted to technology. Speaking realistically, purely from the viewpoint of technology, most of Asia cannot compete with the West in even the most basic scientific capabilities. Add to this the fact that under actual market conditions, developing cutting-edge technology requires huge amounts of capital and the assumption of high risk. The U.S. has the capabilities to do this, but most Asian nations do not. Even if Taiwan, for example, were able to be at the world's cutting edge, we would not be able to create value by commercializing and marketing the technology with the speed that the U.S. can achieve.

Companies undoubtedly should pursue innovation in leading-edge technologies, but this is not the most pressing area of concern. The most urgent matter is using technology to pursue innovation that creates value for consumers and the marketplace. If this is done well, not only can costs be lowered, but the contribution to humanity at large is greater. Even though the U.S. may be the ultimate source of most technological innovations, Asian companies play a key role as well. That's because when American companies discover that the cost of creating practical applications of technology is too high, Asian companies can take over at that point and add their own innovations in the process. Why doesn't Asia in general position itself in this role?

Moreover, innovation is possible in any sphere, whether in business practices, technology, products, marketing, services, or the supply chain. The key is in whether something is new, and in stopping oneself from just doing "me-too" things. If a product's industrial design gives one the feeling that it's been done before, it should be rejected. Avoid at all costs the hive mentality because just doing the same thing as everybody else is not what we should be doing—this resolve is absolutely essential.

CREATING VALUE THROUGH BUSINESS PRACTICE INNOVATION

There are many types of innovation in business practices. Some use premiums to increase market share, with gradual returns in the form of future profits. For example, shaving razor and camera vendors use this form of innovation. The creation of new lines of business is another way to innovate because it gives rise to a new operational mode. The same product can be the basis for a completely new business: Dell Computer is a well-known example of this, and Taiwan Semiconductor Manufacturing Corporation (TSMC) is another company that created a huge new line of business (foundry services for semiconductor designers) even though no "new" product was created.

These two companies both created new ways of doing business and new value for customers. As most people know, Dell broke from the traditional model of computer sales, where a computer passes through several levels of distribution before getting into the hands of consumers, and instead uses direct sales. TSMC serves as a chip foundry for companies without the budget or expertise to do their own manufacturing, allowing them to concentrate on chip design. This model provides fast speed, low cost, and low risk, and it reduces the time needed to commercialize a product.

The creation of these business models gave Dell and TSMC an advantage. There are many companies that have been forced to change their business practices because their original business models turned out to be dead ends. In the future, cases like these will become more and more common. Even for new industries, after just five or ten years, the rapid development and strong profitability that existed at first may not be sustained with the original business model. In order to survive, the business model will have to be changed.

Dell's direct sales model has forced the issue in the PC industry by turning the retail channel business into a money-losing proposition. Four or five years ago, it was noted that Acer could not afford to continue losing money in its U.S. operations, so we got out of the retail market. The reason was that the business model in the channel could not be used for selling PCs. I wanted to change this business model, and I sought out many different companies, including Intel, to form alliances and persuade the retail channel to change their thinking about the channel. Perhaps it was a matter of too little force applied to too large a stone, but the situation could not be changed. Since the odds were stacked against us, Acer decided to simplify its operations and get out of the retail market—in any case, it wasn't going to kill us.

Creating value through technological innovation

For the smaller Asian economies to continue their development, technological innovation is very important. To create a new business model, these markets are too small in the sense that an approach that works in such markets will not necessarily be successful in a large market. Even if a company attains a leading position in a small market, when it enters a large market and a big company gets aggressive, the odds are stacked heavily against it being able to duplicate its original success.

Business models are more localized, but technology is global, meaning that if you get the technology right your opportunities are unlimited. For example, in 1986, Acer launched a 32-bit PC ahead of IBM. Putting aside the benefit in building a reputation for innovation, we also won a lot of orders as a result and earned high profits. In the area of high technology, Intel's CPUs are a great innovation. Intel took advantage of the opportunity presented by IBM-compatible PCs and established x86 microprocessors as an industry standard. Moreover, Intel uses constant innovation to improve its price/performance ratio, raise competitiveness, and establish product recognition.

Creating value through product innovation

Meeting a customer's special need, creating a product that fits a particular lifestyle, or making a product more intelligent—these are all examples of product innovation. Blue jeans can be tailor-made, and later with the customer's basic information available, new styles that fit the customer perfectly can be offered at any time. In some shoe stores, foot sizes can be measured, and all the data entered into a computer. Customers will never need to remember their shoe size again!

As for lifestyle products, precedents are even more numerous. Besides Apple's iMac, Swatch wristwatches are a successful example. The Swiss clock industry was put under severe pressure by its counterpart in Japan, so it came up with lifestyle products like Swatch. A person can have many watches to suit his or her lifestyle. Right now, the hottest mobile phones are also products that fit in with particular lifestyles. Another route to product innovation is creating "smart" products that can "think"; for example, Sony's Aibo robot dog. Otis elevators are more complicated— the company has placed many sensors inside their elevators so that if there's any kind of problem, headquarters is notified directly through computers and can dispatch people to make repairs immediately. Product innovation such as this provides much room for different approaches.

CREATING VALUE THROUGH MARKETING INNOVATION

When Intel began heavily promoting "Intel Inside" and sought me out to participate in the marketing campaign, I didn't yet have a sense of its power. I didn't understand why Intel would want to communicate directly with consumers because not only would it cost money, but the people who bought Intel products were not consumers but companies like Acer. Later, Intel pressured suppliers to jointly advertise with it, partly subsidizing the suppliers' cost but demanding that the Intel Inside musical tones be played in any television ad. Regardless of which company paid for the ad, that sound would always be there. Intel even stipulated that in all print ads, its logo could not be smaller than the supplier's, and had to be placed in a prominent position. Very obviously, it was using all its customers' advertising collectively to advertise itself. It used this form of innovative marketing to create maximum advantage for itself. Airlines' frequent-flyer plans are another marketing tactic designed to build customer loyalty.

Asian business needs most to strengthen its ability to innovate in marketing. However, it will be an enormous challenge to build up the required internationally minded talent because of the geographical and cultural distance that separates Asia from most of the world's major markets. When the company changed its name to Acer in 1987, annual revenues were around US$400 million (at present, they are around US$9.4 billion), and though we were small, we were ambitious. We went to Ogilvy & Mather in the hope of using a limited budget to train marketing people who could have "a view of the world, but from Taiwan's perspective." At the time, "marketing people" referred only to advertising creative directors, but the plan still failed. Later, I sought out the world's largest advertising company to jointly fund a new company with the aim of using limited expenditures to cultivate marketing talent in this part of Asia, but again the plan fell through. The real key is people and experience.

CREATING VALUE THROUGH SERVICE INNOVATION

The strength of the Internet is its accessibility from any place and at any time. The development of banks was based on this kind of accessibility as well, from branch ATMs to inter-bank cash withdrawals, and finally to online banking. The best-known example of how to give customers peace of mind is FedEx. Its customers can at any time query the present location of any parcel. The Internet can be used to understand customer inclinations through such means as membership clubs; this allows you to know which members like what things. For example, after buying books online, the Internet bookstores will recommend books to you, with Amazon.com being the best example.

CREATING VALUE THROUGH SUPPLY CHAIN INNOVATION

Acer and Wal-Mart, which is often considered the most influential retail business in the U.S., have had several cooperative ventures, but none of them have succeeded, with the main reason being that Wal-Mart's supply chain management system does not excel at handling "fresh" products. If the operational models of supermarkets and large retailers could be integrated, an entirely new kind of business may be possible.

At present, the production of computers needs to be consolidated, and yet the finished products must be the freshest; they must not be left in the store to "spoil" but rather should be delivered straight into the hands of customers. Dell resolved this problem, but up to now its products target people who are familiar with computers. For first-time computer users, Dell's direct sales are less appropriate. Acer's Aspire PC attracted many first-time computer buyers because it was simple and attractively styled. Selling computers through retail outlets provides low profits, with a high service burden and a high level of inventory. The risks associated with price-cutting are also high, so it's evident that this is not a good distribution channel. In the Internet age, the Web is the basic infrastructure for information exchange, and it provides financial transactions, data retrieval, and services. If it finally becomes a major channel for distributing physical goods, quite obviously the roles of all the players in the supply chain will have to be adjusted.

CONCLUSIONS

To sustain a business over the long term, and to constantly enhance competitiveness, the most important item in the competitiveness formula is innovalue. Using Acer Aspire as an example, it was launched in Taiwan and the U.S. simultaneously in 1996, and it definitely provided excellent value because it was innovative and carried a very good image. Compared with the situation in Taiwan, service in the U.S. was poor and product quality was less consistent. As a result, the value created in the U.S. was lower. From the point of view of material costs, the U.S. and Taiwan are the same. However, compared to other products with the same level of functionality, the Acer Aspire came in a different color, the display had a different casing, the control panel for the CD-ROM drive was different, with the end result being that expenses increased while flexibility decreased, adding to the cost of intangibles. Acer Aspire created value in the U.S. but raised costs, so there was no obvious increase in overall competitiveness. However, in Taiwan, while value was created, costs were well controlled. The result was that the competitiveness of Acer Aspire in Taiwan greatly increased.

Customer values are constantly shifting because customers attracted by the new easily disdain the old, making the view of a product's value change over time. Therefore, you must have a sufficient understanding of what the customer is thinking. Acer has always stressed the message "we hear you," conveying the idea that Acer listens to customers and wants to understand their needs. This became a very important capability that contributed greatly to the development of Acer as a business. If you want to innovate, don't pass up any opportunity because innovation can sometimes be simpler than you imagine and just a matter of willingness.

Innovation is not just a matter of extravagant fancies, but should be grounded in basic knowledge. You need to be aware of many issues because once you conceive of an innovation, you must use the competitiveness formula to immediately assess the costs that the innovation carries. From the perspective of management and corporate development, innovation is the business activity with the highest return on investment. If you can cultivate the habit of innovating, it gradually becomes easy.

DISCUSSION

Q: **What indicators can be used to distinguish whether a business is innovative, customer-centric, or has core competitiveness?**

A: Comparisons should be made constantly. A free economy is all about competition, so comparisons should be made to industry rivals, to bellwether companies, or to your own business in the past. As for what quantitative indicators should be used to make comparisons, in fact quantification is not necessarily the most scientific approach. That's because many factors, such as customer satisfaction, are difficult to quantify. Aspire Academy once invited people to talk with our executive staff and discovered that customer satisfaction was expressed in terms of intangibles. It is not a problem of product quality as customers have an immediate sense of whether product quality is good or not, and this judgment is not necessarily based on visible factors. The most practical measure is to rely on feeling. You can make comparisons with industry peers or with your own company in the past; of course, you cannot do this alone, but must seek out several people to give their sense of things, and make comparisons in one area after the other. For example, if core competitiveness is based on low manufacturing costs, then compare in this area: Do you have the lowest costs? What is the advantage of having low costs? If just having low costs has no impact on competitiveness, or if the part that low costs play in deciding success or failure is very minor, then it's not useful.

Understanding customers is a prerequisite for creating value

Q: **Innovalue is judged by consumers, and consumers in different regions have different needs. How can businesses get closer to consumers and understand their lifestyles?**

A: It's essential to understand customers, but since understanding customers carries a cost, the methods to use become an important issue. You may perhaps reflect on experience to come to some conclusions, and this approach carries the lowest cost. If you gain a true understanding, you have taken the first step toward creating value.

However, it's also possible that much money has been spent to no avail—for example, because a questionnaire was poorly designed or because the survey didn't reach target consumers—with the result that there is no way to get the answers you seek. Carrying out a survey is similar to advertising: you must make presentations. The people who specialize in conducting surveys, like those in advertising, and those who specialize in products, may be far removed in their areas of expertise. If you don't have good communication and misunderstandings develop, the extra costs will be very high.

The capabilities of an organization, and its ability to learn, must be enhanced; the question is whether an organization can possess this kind of system to help everyone within it grow. For example, we have worked with Japanese computer companies for a long time. After a few years, the Japanese had come to understand that Taiwanese companies were willing to introduce products on to the market even if their product quality was not up to the highest standard. Many Japanese companies take product quality to the point of perfection before releasing a product to the market; although they have mastered product quality, they have missed the timing, so the value they create is thereby greatly compromised.

Have a grasp of consumers' core values

From Acer's point of view, as the scope of a business expands, these aims gradually enter a communal consciousness and understanding: deciding on the most effective approach for grasping the thinking of customers and setting the necessary objectives. For example, we ask: Who are we making computers for? Selling to OEM customers is very simple, but selling to students is fruitless because most of them are do-it-yourselfers and probably won't be interested in buying your product. When we went to sell to large enterprises in Europe or the U.S., they asked, "Who is Acer?" After hearing that question, we knew we could not win.

If you do not ask whom you are selling computers to, it might seem that anyone will do, but after some real analysis, a target customer group will emerge. After verifying who your target consumers are, things become much simpler—all you need to do is to pay attention to what this group of customers cares about.

Acer has three types of customers: consumers, businesses, and other companies in our industry. What do they want? What consumers want is simplicity, what businesses want is reliability, and what other companies in the industry want is a partnership. They may want a low price, a high-quality product, or high performance, and so forth, but they're not usually decisive. Once a target is set, thought can return to the design, manufacturing, and sales processes, and on to after-sales service. Although we have our core competitive strengths, we must also know where customers' core values are situated. The ideal is when the two are combined.

Corporate leaders should train themselves to be able to look at an event and quickly grasp its essential meaning. Doing business is like doing battle, where you must be able to use weapons and tactics as if they were second nature because in the heat of an actual battle, you don't have time to think up new tricks. So you need to be able to think of the most effective approach, understand what the key issue is, and then take action right away—these are the capabilities that we must have in facing a complicated business environment.

The globalization of trends and functionality

Q: What do you believe globalization really is? From the point of view of globalization and market innovation, what did Acer Aspire signify, and what did Acer learn from its experience?

A: We see two types of globalization. The first is a trend. Because large companies are so numerous, competition has become globalized. Whether it is capital, products, technology, or people, everything can be circulated freely worldwide. When you are an individual or an organization, the way to deal with the globalization trend is to use approaches that recognize the implications of globalization. These approaches include the understanding of markets and of technology because you cannot isolate yourself from changes taking place in the rest of the world.

In addition, you must turn ideals into actual practice in your entire organization; this means that an organization which is beginning to internationalize must internationalize every part of itself and its operations. For example, going from Taiwan to Hong Kong is internationalizing, although the significance is not as broad as globalization. If we say that Hong Kong represents South-east Asia, when we set up in Hong Kong we have become international. Later, if we go to Europe to set up an operation, or to the U.S., and to Australia, globalization slowly takes shape. Once the process of globalization is more advanced, you find that Hong Kong cannot represent South-east Asia, and you must penetrate more deeply.

The second type of globalization is in business functions. For example, at first you may only do manufacturing, but later you may start to do marketing, R&D, and even financial management. Although most Japanese companies are globalized, their R&D and financial management are Japanese, and their approaches are not globalized.

Innovation can be created with little effort

A definition for innovation needs to be formulated—is innovation just the creating of something that's different? For example, let's say Acer takes U.S. approaches and introduces them to this part of Asia. Because the business environment and market here are different, we need to make small adjustments so that they are just right for the market. Innovation can in fact be the product of a little effort. Don't just do what everyone else is doing, or just imitate someone else's successful formula without making necessary adjustments—it just won't work.

Innovation can be classified into different grades, with the higher the grade the higher the risks it carries, but the higher the potential rewards. Companies with a large home market are better suited to pursuing high grades of innovation. Local companies pursuing the U.S. model of breakthrough innovation should not be viewed approvingly, and the reason is simple: the local market is too small, and the cost of this style of innovation too high. The risks are great, but the value is limited.

What we are pursuing is innovation in the context of the business conditions under which we are operating. In the context of the global environment, our level of innovation may only be at an intermediate level, which I described as the "number two" concept. At the time Asia could not see beyond "number three" or "number four," Acer was aggressively seeking out a place as a world "number two." Some people criticized us for not aiming to be "number one." But the fact was at that particular point in time we didn't have the qualifications necessary to be the world's number one, and the issue could not be forced. The only realistic approach was to wait for the conditions to mature and then see. Currently, Acer has already achieved the status of being number one among "number twos" (in this part of Asia, we are number one), and we are progressing toward becoming a partial number one.

Pay attention to the basic capabilities established by pioneers

Q: Taiwan's electronics and information technology industry is number one in many product categories, such as monitors, scanners, and so forth. However, these products all seem to have served out their usefulness on the market. What recommendations do you have about these products?

A: Taiwan's monitor products depend on the same basic capabilities as producing television sets; however, Taiwan's television and home appliance companies did not succeed internationally. Our initial forays into the international market were not inferior to Korea's, for example, Tatung, Sampo, and Teco all entered the U.S. market, but because the scale was small, we didn't make any inroads and were not in the end successful. Still, this industry established the foundation we could build on with monitors.

Therefore, we must not neglect the basic capabilities established by such pioneers. Whether from the perspective of industrial design, technology, or mass production, Taiwan already possesses some basic capabilities. The biggest problem now is that even if the volume is significant, it's still not valuable; every year profits decrease, and a company that could in the past support 500 people can now only support 400. What we should be thankful for is that to do the same things as in the past, we may not require 400 people; so if you have an aggressive plan of action, perhaps just 350 people can get the job done.

Before a competitive advantage disappears, prepare for future eventualities

With these things in mind, what should experienced managers do? Acer put forth the concept of specific-use computers several years ago, and before that the application-specific computer, and has continued to direct efforts in that direction. At the time, I said that we should take advantage of the competitive strengths we had already built up to seek out new opportunities. Later, I discovered that the idea was not workable because a specific-use computer was only hardware—without the appropriate services and software, it was all just empty talk. So Acer got aggressively involved in developing X-services, with the hope of exploiting the foundations established in the past by the company or local industry.

At one point, Acer lost US$200 million in the U.S. market during one year. I told our people that in doing R&D, the process itself is the most important. Even if your plan fails, there is certain to be valuable technologies, or at the very least development experience, that can be applied to future efforts.

By the same token, Acer's establishment of its brand in the U.S. and the world carried a very high cost, though fortunately two recent events have given me a certain amount of consolation. One of them was Acer's entry into the computer peripherals business two or three years ago. Even though brands in the PC business are essentially worthless in the sense that they cannot help their owners earn a profit, when you turn to peripherals products, there are few companies with a stronger brand than Acer, and this is why we have been able to make money with peripherals.

The strategy was to attack at the peripheries; if complete systems didn't earn money, then do peripherals. Additionally, we invested in startups and earned back enough to make up for our earlier losses. In sum, we must prepare for a rainy day while we still have our core competitive strengths and before our advantages disappear.

Allocate your limited resources in areas of greatest competitiveness

As for Taiwan's gradually disappearing peripherals vendors, I think the solution is simple: get out of the business and get into mergers.

The reason that the U.S. competitiveness is the greatest is very simple: it has allocated resources to higher value-added industries. From the point of view of national competitiveness, what is the point of using up so many resources to produce products without any added value?

Ten years ago, Acer developed telephone modem products, but later transferred the business to another company in the industry and let it try to make money from the venture. The reason for this was that if Acer couldn't create a successful business around that type of product, it was just a waste of resources. Today, if many local industries want to maintain their position as a world number one, they should direct their limited resources to areas in which they are most competitive. It should be stressed that in many Asian nations where resources are extremely limited, companies must make the most of the resources, like managerial or engineering talent, and direct their efforts to the areas where they are competitive and have the opportunity to win.

Q: **E-commerce is beginning to take off, and there is talk everywhere of B2C transactions. Why did Acer decide to team up with Taiwan's On King electronics chain stores in March 2000? Was it because you didn't like the prospects of B2C e-commerce?**

A: In terms of maturity and volume, B2B e-commerce in the U.S. is far ahead of B2C. If we're talking about B2B, we must consider the fact that we are part of the world community. B2B in Taiwan's economy is already internationalized, and the government is promoting many initiatives. We should not only get involved in upstream, midstream, and downstream relationships locally, but do the same globally.

B2C can be divided into the sale of physical goods and the sale of intangibles, and according to whether after-sales service is required. After buying physical goods, consumers may need further service, and that has to be delivered in the real world. The alliance between Acer and On King was created to effectively combine the real and the online worlds.

Virtual dream teams

I also came up with the "virtual dream team" concept. The team in this case refers to a combination in which each entity involved has its specific role, just as in a team sport, where the player in each position has his or her particular role to play. The Internet has a role, as does On King and each individual person.

The present era is one of extreme diversity; we no longer just sell one kind of merchandise or one type of service. In order to reach a certain objective, we must classify tasks, with a different team used for each type or instance of supplier/consumer relationship. This team includes the supply chain, banking, shipping company—in the final analysis, what is the key element that must be carefully coordinated? It's just like a team sport, which requires a closely coordinated group effort.

Therefore, the Acer Group participates in many different business initiatives. For a specific objective, we quickly formed a virtual dream team with the core competencies needed. For a business to establish a new core competence is not easy, and even more so in the case of completely new areas that no one has attempted before. Somebody else may have already developed the core competence that you need, and at that point you need to decide whether to try and create the core competence yourself or assemble a virtual dream team to meet the objective. With this issue in mind, we hope that we and On King can become a virtual dream team; however, whether we can utilize the management know-how from the past and the technology from our established product lines or new products to help On King to more effectively develop its current business is still an open question.

Establishing international marketing capabilities

Q: **As you just said, marketing talent in Taiwan is extremely scarce. Could this become a crisis for Taiwan, or more broadly, for Asia?**

A: In developing long-term competitiveness, establishing marketing capabilities is the most difficult and requires the most time. Taiwan's economic development began with manufacturing, and doing manufacturing for foreign companies helped hone our managerial capabilities. Next, skilled people were trained in Taiwan, and some others returned from abroad, gradually allowing Taiwan to build up further capabilities—and this required time.

Taiwan's manufacturing capabilities rate an A, its R&D capabilities a B or B+, and its marketing capabilities a C or D. We have also tried to learn from marketing experiences from abroad, but it's not global in scale. When managers and advertising companies take Taiwan as the target market, it is a simpler task than trying to market to the entire world. Not only is the scope small, but the depth is inadequate. It's entirely possible that it could take fifty years to establish international marketing capabilities in Taiwan.

Only by having international marketing capabilities can companies fight truly global battles. Taiwan must accumulate experience, and we must not only consider the quality but also the quantity of marketing resources because only when quantity reaches a critical mass can there be a demonstrably favorable environment. Under these circumstances, it's not worthwhile to fret about the current weakness in the area of international marketing, but we must be aware that if we hope to raise our future competitiveness, it will mean rectifying our current shortcomings. Moreover, if this process of improvement is stretched out over time, it's a manageable investment burden.

Avoid unrealistic expectations

In doing international marketing, we must avoid having unrealistic expectations. Even though Acer has been very aggressive, we also make adjustments depending on the circumstances. However, we will definitely not give up and will keep pressing forward step by step. In fighting this battle, we have time—that is to say, Taiwan's high-tech industry has in fact enough time because outside of manufacturing there are many areas in need of strengthening.

First, let's not speak globally, but start with Asia. Excepting the Japanese and Korean markets, we should be able to take the lead. Acer is number one in South-east Asia; according to a reader survey conducted by Reader's Digest, in this region Acer is ranked ahead of IBM and Compaq, demonstrating our strength. Looking at Europe, many companies' domestic sales account for only 50% of total revenues, intra-Europe sales for 80%, and no sales to Asia. So we must make Asia our focus. We should do this because in Asia we can compete on a fair basis; when U.S. companies come to Asia, they are not only foreign companies but also far away from home. We can get what we need regionally; in Europe, we are fighting a distant battle, and our chances are not that good.

As for workforce training, the most important thing is that if the company can handle the financial burden, some meaningful actions should be taken and global ambitions given freer reign. In training people to push a company on to the international stage, though the experience of other countries can be considered, just relying on this information is useless—the information must be digested and made to suit our particular circumstances so that it can actually be applied.

Innovation must be handled carefully

Q: Does the ability to innovate imply profitability? Will a good new product ensure profits? How can this be assessed?

A: It still comes back to the competitiveness formula. Innovation does not by itself guarantee the ability to turn a profit—the ability to execute and the right timing are also needed. Some innovations may not immediately produce value by themselves, but can stimulate an atmosphere of innovation and produce something valuable for society. Value cannot be measured only in terms of money; a positive influence on society is also a type of value. In science, much innovation just forms a resource that others can apply.

Regardless of how well a product is received by the market, there is always an element of risk. Will the process of innovation go as smoothly as you imagine? Only time will tell. However, looking at things from the other side, if a company does not innovate, but just tries to milk whatever it was doing originally, is there still a chance to survive? The answer is that if you do not innovate, you are as good as dead; you're just living on borrowed time. So if innovation also means death, well, if you've got to go, you might as well go out with a bang.

Naturally, innovation must be handled carefully, and it's mainly about creating value. If your workforce is not accustomed to innovating, or can't do it, it will add to costs and lower performance. You will have to return to the formula, evaluating where the key factor is. The greatest challenge is to choose, out of a mass of options, two or three factors—ignoring the rest—to consider and then make a decision. If by chance things do not go smoothly, at least you know what factors were neglected.

From another perspective, Acer created much value through its management innovations. This value was created from scratch, and it helped develop the potential of our people. We created an environment in which workers could fully utilize their skills, so management innovation is definitely a way to create value.

CHAPTER 6
Eliminating the Obstacles to Innovation

To run an international business requires an environment for nurturing innovation, and the ultimate responsibility for creating such an environment rests on a company's leaders. The way a leader of a unit or organization conducts a meeting, or the way that he or she speaks, can have a profound influence on the organization's ability to innovate.

Factors that influence innovation

There are both external and internal factors that influence the environment for innovation. External factors include market scale, industry conditions, capital markets, protection of intellectual property, and social culture. Internal factors are corporate culture, organizational structure, and a system of incentives, as well as whether the company provides an environment for personnel development that allows people to constantly learn—these all influence the ability to innovate. The most important factor is the personal style of the company's leader.

Market scale is the driving force of innovation

The market is the driving force behind innovation because the market provides incentives for people to innovate. The U.S. is the world's largest market, and it has gathered talent and wealth from all over the world, all seeking innovation, and it is where you can achieve the greatest returns. Mainland China's market is potentially as large as that of the U.S., and the rest of Asia should make the most of it. Large markets attract people to take the risk of failure of innovations because if you do succeed it means considerable returns. Moreover, if a market is large, the same formula for success can be repeatedly used.

The local Asian software industry is very unlucky in that even if it develops some promising software, the volume of sales it can attain in its home market is very limited. But in the U.S., the potential volume is 100 times greater, and the difference that makes is like day and night. Moreover, once you achieve success in a market, you must constantly repeat the success with the same formula in that same market because to try to duplicate it in another market means dealing with the possibility that its needs may differ, in which case the effectiveness will be greatly reduced.

Most Asian markets are relatively small, and this makes it difficult to justify risks. Therefore, many venture capitalists in the region will not invest in new innovations—the risk is too high. These venture capitalists would typically rather invest in relatively mature products, where many Asian companies can demonstrate their strengths in cost control or task allocation. On the other hand, venture capitalists in Silicon Valley are not willing to invest in well-established industries, and this is a big difference. Relatively speaking, companies in the U.S. can fluctuate greatly in their fortunes. A successful company may be gone in just a few years. Because competition in North-east Asia is less, and business expenses are in general lower, local businesses have greater stamina. This stamina has its advantages, but there are many businesses here that are in a vegetative state—they may be technically alive, but their existence has lost any meaning. In consideration of overall social resources, such "comatose businesses" should be discouraged.

Additionally, when a market is large it naturally gives rise to much competition, and competition is very critical. Why is the level of performance in so many sports so high in the U.S., for example, in basketball and golf? The reason is very simple—it's because the market is large. Only through constant competition can this result be produced. The best talents in the world are all concentrated in the U.S., and it can accommodate the investment of even more resources. My hope is that mainland China can become the engine for innovation by Chinese, and that hope is founded on the basis of its enormous potential market.

INDUSTRY INFRASTRUCTURE FOR INNOVATION

Industry infrastructure and local industry clustering are definitely related to innovation. If an industry is extremely competitive, then it is necessary to constantly innovate in order to keep up with rivals. For example, in the past, Japan was extremely competitive in consumer electronics products, with competition much more intense domestically than abroad, resulting in a constant stream of innovation.

In addition, industry clustering, such as the Hsinchu Science-Based Industrial Park or Silicon Valley, can also speed up the pace of innovation. Once such a clustering is established, working within its disintegrated structure can also allow an individual business to concentrate its capabilities on a certain task and share risks with other companies. If there are risks associated with a larger initiative, tasks can be appropriately allocated and coordinated among upstream, midstream, and downstream satellite businesses; and if by chance the direction is wrong, everyone involved can communicate and the plan adjusted quickly. Even if some losses are incurred, they are shared by all the parties involved, and the loss for individual companies minimized. With this concept of shared risks, innovation is easier to realize. The notion of venture capital is based on just this idea of shared risks.

Also, if the structure of an industry is incomplete, then more time must be spent in order to achieve the most basic critical scale. Therefore, it is difficult in these circumstances to make further attempts, and the chances for success are small. If a region's industrial structure is mature, it is easy to attain a critical mass of economic scale, and with this scale, development can proceed in a snowballing fashion.

For example, to pioneer a new market in the U.S. might require spending ten million dollars to even have a chance. However, in Northeast Asia, because of its small market, a company might need to spend just one-tenth of what its U.S. counterpart would. Because of the special character of Taiwan's industrial structure, spending just fifty thousand dollars will allow you to reach the same cost and economic scale as in the U.S., and this innovation is worth doing because you can follow up by then seeking opportunities in the international market.

An overwhelming majority of stronger products in this part of Asia, such as notebook computers, computer peripherals, information appliances, are very competitive because they take advantage of the uniquely Asian assets of low cost of engineering, industry clustering, and fast speed, added to the high concentration of satellite vendors, naturally giving rise to the needed economic scale. The economic scale needed by individual companies is small, efficiency is high, and risks correspondingly low.

THE INFLUENCE OF CAPITAL MARKETS ON INNOVATION

Capital markets are also an important force behind innovation. Some innovative ideas may not be taken seriously in the organizations where they originated, or they may be difficult to put into practice. There are also people with good creative ideas who do not want to let their company know and want to keep them for themselves (this should not be encouraged!). These innovative ideas may then become the basis for a new company.

Experience in managing venture capital investments can be recycled. Any venture capital company manages dozens of companies, and a large one may manage hundreds. In the process of building something from nothing, a venture capital company will generally get involved in much management, and during the handling of each case it can provide the appropriate services, and so the chances of success are higher. Venture capital funds not only help in the starting up of a new company, but also use the money from successful startups to help newer ideas.

The development of Taiwan's stock market in the late 1980s played a critical role in making possible the current success of Taiwan's high-tech industry. Because capital markets are critical to business development, government policy in Taiwan encouraged companies to go public. Taking Acer as an example, we did not originally intend to go public, and only did so after being persuaded. The vigorous development of Taiwan's capital markets has given high-tech companies access to plentiful capital. This naturally makes these companies willing to press forward aggressively, particularly in view of the relatively low level of risk they must shoulder.

Furthermore, the concurrent development of the capital markets and the high-tech industry was the key factor that allowed Taiwan to largely escape the effects of the Asian economic crisis. First, because the high-tech industry was highly competitive, its added value was also high. Then, the vigorous capital markets allowed Taiwanese companies to control a large proportion of their capital base, making them less vulnerable to the effects of the currency crisis. Most companies in South-east Asian countries, as well as in Japan and Korea, control a relatively small proportion of their capital base. On average, Taiwanese companies controlled 50 to 60% of their own capital base; but the proportion for Korean, Japanese, and South-east Asian companies was only 20 or 30%, or even as low as 10%.

THE INFLUENCE OF SOCIAL CULTURE ON INNOVATION

Whether a society protects intellectual property rights is also a key factor in whether industry is able to innovate. This is because the results of innovation usually are a form of intellectual property—and the question is whether society respects its value. Additionally, another key indicator of how strongly innovation is supported is whether the social environment encourages an entrepreneurial spirit. If many people are starting up businesses, this will naturally give rise to an environment for innovation. Innovation carries many risks and requires collective understanding and communication. Whether there is adequate communication when policy decisions are made will also have an influence on innovation—this is an issue at the level of social culture. Although innovation originates with the individual, when it comes to actually instituting an innovation, because it may not be fully formed and may change at any time, many policies must be adjusted accordingly. If this process is not done in a transparent fashion, it is very difficult to persuade everyone to shoulder the risks associated with introducing an innovation.

In the early stages of running Acer, I would always talk about risk during interviews with prospective employees. At the time, Acer was involved in selling microprocessors, and workers were candidly told that I didn't know if the company could survive because it was taking the path of innovation. However, I did encourage them, telling them that even if the company went down, all the experience they gained would prove very useful because we were involved with technology that would be pivotal in the future. Had things not have been made clear to employees, they would not have been willing to innovate along with me. Innovation is not about wild fancies, but about breakthroughs achieved on a foundation of comprehensive knowledge. Innovation means already having assimilated considerable experience, and only then searching for the breakthrough, seeking a new approach.

Social culture can also be a negative force, impeding innovation. If the person who is the driving force behind a project controls things too tightly, it is difficult to innovate. This is why it is more difficult for traditional family businesses, or companies with a very top-down organizational structure, to innovate.

The educational environment is another pivotal factor determining the level of innovation that can be achieved. When speaking of education, businesses should think of it in the broadest sense, with schools being the least important educational institutions. All that the schools can do is provide basic learning and cultivate the habit of using knowledge. Even an educational system based on rote memorization, as is common around Asia, cannot kill innovation, unless even outside of the schools, in the workplace, homes, and organizations, thinking is restricted and people are kept from having new ideas. So, in fact, the "education" provided by other organizations and actual social conditions is more important.

Finally, innovation requires a leader—and within a country this ultimately means the government. Unfortunately, as we all know, governments are not by their nature very innovative. If a bureaucrat tries to be even a little innovative, he or she is likely to meet with obstacles to implementing the idea. In any case, if the value of new ideas is denied without thought in a society, this kind of society's values and environment are unfavorable for innovation.

THE INFLUENCE OF CORPORATE CULTURE ON INNOVATION

The most critical of the internal factors affecting innovation is corporate culture—whether or not a company encourages empowerment, allowing workers to make decisions independently and tolerating some mistakes. At Acer, we call the creation of an internal environment for innovation "paying the tuition for workers" because the company is willing to make a sacrifice in order to give employees the opportunity to learn how to innovate.

Democracy is ubiquitous in the most innovative companies—whether people are allowed to freely express their ideas in meetings is a big test of how deeply rooted democracy is in a corporate culture. Many people talk about the open exchange of ideas, but are the conclusions that a business arrives at produced through the refinement of ideas freely expressed in a number of meetings? This is not only a matter of innovation; the conclusion may not be the leader's original idea, but arrived at through a process of group brainstorming and discussion. To make the implementation of an innovative plan more effective, putting the concept of the open exchange of ideas into practice is very important.

In many Asian companies, meetings are held in order to let the boss hear what he or she wants to hear, but even if a worker wants to be supportive of the boss, he or she still must speak the truth. Because having different positions within a company gives people different perspectives, employees' understanding will differ from that of the boss, and these differing views must be expressed. Can the leader accept such differing views? This is a matter of leadership—a leader should encourage anyone who is able to offer a new perspective and not adopt an authoritarian style.

Anyone who cannot come up with any new ideas or viewpoints, but who only follows the crowd, has no growth potential. During meetings, many views can be very constructive, and we can assimilate these and immediately make adjustments to our way of thinking. There will also be destructive suggestions, and we must not be angered by them, but instead seek opportunities to communicate and explain. Even in handling the big initiatives that require long-term planning, top executives do not necessarily have to attend every meeting. For example, when we were planning the e-Life Show to showcase digital and Internet technologies to the general public, Acer Group CEOs and I participated in only a few of the countless meetings that were needed. And of these few, most were just me communicating directly with employees having relatively low rank within the organization. I used these gatherings to convey Acer's meeting culture, and not turning them into one-man shows.

In doing things in this manner, a conflict between discipline and innovation may arise. Some people use maintaining discipline as an excuse for suppressing innovation, and leaders should face this problem squarely. Innovation must have a base in knowledge and must be sought while keeping some basic principles in mind. Otherwise, things will just degenerate into idle fantasies. Innovation *must be* built on a base of discipline, which gives it more value. The basic rules for an organization must be spelled out very clearly. Taking the Internet as an example, its network protocols are the rules, and any innovation in which they are not followed makes communication impossible and therefore has no value. Rules mean making basic demands on an organization and individuals—an eight-hour workday is a basic rule, and how to make work hours more flexible is an area where there is room for innovation. There does not have to be a conflict between innovation and the order created by rules.

THE INFLUENCE OF ORGANIZATIONAL AND PERSONNEL RESOURCE DEVELOPMENT ON INNOVATION

Personnel development in an organization is a big issue. Compared to other organizational forms, an internet organization is more effective because it does not represent an organizational structure built on a "might is right" mentality. An internet organization is like a network—it easily encourages innovation, and one event can inspire new ideas and may even have a broader effect. Through an internet organization, everyone can make contributions to innovation at the same time.

Virtual teams can also be a means of facilitating innovation. In order to confirm the viability of an innovative idea, anyone with a stake in this innovation or its objectives can at any time form a virtual team to speculate and brainstorm, with members stimulating each other's ideas. However, this brainstorming must be grounded in a very solid base of knowledge.

Leaders are more important in learning organizations. Note the plural "leaders"—the managers of each department are all leaders. They must decide on how to train people within the organization, and after an innovation is implemented they must make participants feel that they share in the glory.

ACER INNOVATIONS

In the early 1990s, Acer ran its first executive training program, training 100 executive-level managers. In the late 1990s, we established another similar program and trained 200 executives. In nine hours of classes, I focused on vision, business philosophy, leadership style, corporate culture—all the things that are very important. Thanks to such programs, the Acer Group possesses a strong pool of managerial talent, which is very important in Asia.

Even more importantly, in Acer's open environment, opportunities to learn are always available. An open environment means not deceiving employees or withholding information, and what an employee sees is not manipulated or packaged. He or she can see the ugly side of the company and learn how to avoid repeating the mistakes of others in the future. When an employee sees the positive things, he or she can learn directly from them. The corporate culture at Acer encourages people to have a broader outlook and to be more pragmatic in their approach. And while Acer's potential for technological innovation may be more limited compared to Silicon Valley companies because of business environment factors, we have tried to compensate by innovating in the area of management.

Technological innovation is global in nature—any new innovation can be used throughout the world. So technology is not only broadly reproducible, it actually encourages imitation. Even though there are appropriate mechanisms for protecting intellectual property, patent rights are time-limited. Once this time expires, the ideas covered by the patent become public assets for the world—so technology in the end belongs to everyone.

In Taiwan and several other countries in Asia, there are limitations deriving from the business environment, market scale, investment practice, and other areas that make it difficult to create cutting-edge technology. For the past twenty years, we have never given up on technological innovation and will continue to dedicate effort to this area. However, in terms of operational efficiency, the returns will be greater by using limited personnel and other resources to focus on managerial innovation. Every region and industry has its own mode of innovation—there is no universal standard that can be applied indiscriminately. Management cannot be globalized because each case and situation is unique, and there is unlimited potential for innovation in this area.

Very soon after Acer was founded, it instituted a plan to obtain both personnel and financial resources by encouraging startups and then allowing all employees to buy into the startups. This method has been adopted by many companies, demonstrating its value, but twenty years ago it was a new concept. An internal system for nurturing startups started to become popular in the U.S. more than ten years ago, but actually implementing such a system may not be easy. This is because the pressures within these organizations are already great, and competition is intense; the need to constantly increase sales is overwhelming and doesn't allow a loss in focus.

Under these circumstances, for a company to set aside some of its people to establish an internal startup, particularly as these people are key people, gives rise to a dilemma. If the company doesn't forge ahead with the startup, the result may be fatal; but if it does, the move may exhaust the company. That's because taking some of your critical people to establish a startup will immediately result in damage to the company's operations. However, not allowing these people to establish the spin-off will mean they will leave to start up their own, entirely independent company. This is a problem that the company will have to face sooner or later.

Acer's second managerial innovation is distributed management, which was also a very unique approach when it was introduced around twenty years ago. Distributed management is now one of the strongest distinguishing features of Acer, but it has to be done right. The idea of distributed management is grounded in a deep appreciation of the fact that without a spirit of innovation, even surviving in the high-tech industry is going to be problematic. At Acer, every worker is encouraged to make decisions for the company as much as possible, and the company must make efforts to give employees opportunities to learn.

As stated earlier in this book, Acer has expended much effort, spent much money, and used itself as a guinea pig to test new ideas—all for the sake of fostering innovation. We, of course, hope for the best possible results and, in the process, try to do better at innovating. The client–server organizational management mode that Acer came up with has stimulated a trend in Asian and even global businesses. Our development of this type of management model was related to local conditions. Because the market in Asia is not large and because of our competitors' willingness to be niche leaders, we had no choice but to adopt a client–server structure. Now, faced with the industry evolution toward super-disintegration, we also need a new organizational structure, so Acer has begun developing an internet organization structure.

As discussed in Chapter 3, the internet organization concept is a new version of the client–server organization idea. Among all the world's businesses, the Acer Group probably instituted a client–server organizational structure earlier than others, and we must forge ahead and set the pattern for the twenty-first century: the internet organization.

STIMULATING CREATIVE THINKING

Since innovation originates from creativity, stimulating creativity is obviously vital, but how can this be accomplished? Contrarian thinking is the key because the way to get furthest from the current direction is to first make a 180-degree turn in your thinking. If you do this today, you may not have to take radical action tomorrow. Hence, use thinking; after all, it costs nothing.

Contrarian technique works as follows: First, think in a direction 180 degrees away from the current direction, and see if there is some answer for the issue at hand. If not, adjust your thought to somewhere intermediate, and slowly look for an answer; perhaps you will find it at 45 degrees. For example, say that a company is considering whether or not to make an acquisition on the assumption that it will make the company number one in a certain market segment. If your thinking is oriented straight ahead, you may consider only whether the goal of taking the number one position in the market is worthwhile, while neglecting the possibility that the acquisition might prove a burden rather than a benefit. Another good example is the U.S. PC market. The straightforward way of thinking is that increasing sales is desirable, but unfortunately this is a case of the opposite: thanks to market conditions there, the greater your sales volume, the greater your losses.

Whether you seek an answer to a difficult problem at 5 degrees, or 10 degrees, from the direction straight ahead, there is little chance of finding a satisfactory answer because if there were one in that limited area, the problem would have already been solved. People generally tend to have the hive mentality and do the same things as everyone else, and therefore they never achieve anything of significance. Contrarian thinking will generally enable you to come up with a different, creative answer, or at least take a first step toward one.

After contrarian thinking, you must rely on collective brainstorming, and at this time you need someone with a great deal of experience to lead the discussion. In most companies, meetings are generally led by executives with veto rights, and people who have the right both to lead and to veto ideas are usually the chief culprits in killing the brainstorming process. This situation should be avoided. The head of the meeting must deal with inexperienced people, or those who just talk nonsense, but this is also an opportunity to understand why employees think the way they do, and it may even be possible to communicate with them. However, before this communication takes place, the executive must first confirm whether an employee is just talking off the top of his or her head, or if what he or she says has substance. In trying to persuade an employee, you may find that it is you who is persuaded. If you can be persuaded by someone else, this means you have absorbed something new. In running a business, persuading someone else is just selling what you know, which has no value; learning something new is more valuable.

For productive brainstorming to take place, there is a precondition, which has been alluded to already—the company must be democratic; that is, truly democratic. True democracy is a very important part of the environment provided by a company. Claiming to be democratic is too easy. A leader must first liberate him- or herself and make everyone believe that he or she can do so, that he or she can think in new ways and can veto his or her own erroneous ideas. Only in this way is there the possibility of accommodating and encouraging other people to do the same thing. A leader must follow basic principles and must have some unchangeable perspectives and ideas, and not simply deny everything about him- or herself. For example, when a leader does something, the starting point should be benefiting other people. When I say "something," I mean "business." Doing business means creating technology or products, and the working principle should be to think about what other people want or need. Any brainstorming that is done with purely selfish motives will in the end be self-defeating.

CONCLUSIONS

The impact of social education on innovation is greater than that of academic education. The enormous North American market is an important engine behind constant technological innovation, but this should not be attributed to a greater capacity for innovation in a country like the U.S., which is much younger than other countries with a long history and traditions. The real reason the U.S. has become the source of so much innovation is its market and environment; it is not attributable to anything inherent in Americans themselves.

With this in mind, we can see that there is a very good chance that the huge potential market of mainland China in the future will enable other Asian countries to become leaders in global innovalue. Even if the Chinese market does eventually fully realize its vast potential and grow into a market as large as that of the U.S., this does not imply that the innovations that we can produce will only be used in less developed areas. We have the opportunity to innovate constantly and to develop these capabilities. The things we create have a chance to lead the world, with even Americans accepting them gladly.

The venture capital system and the capital markets in Taiwan were the key reasons that it was able to avoid most of the effects of the Asian economic crisis. One of the sources of the Asian economic crisis was the so-called Asian values. The importance of the family is one the foundations of Asian values, and in the business world it is commonly believed that a company is a personal or familial possession. This kind of value, when it exists in a non-transparent environment lacking in rules, lead to company owners thinking of the company's money and their own as one and the same thing. Most Americans have a completely different attitude in that if they start up a business today, it's fine to sell it tomorrow as long as they make money. They think of the company more pragmatically and less personally.

Asian values also imply obedience to authority, and this is very unfavorable to innovation. Thus, creating a kind of intangible, soft, innovative culture will be the key to Asia's future competitiveness. What is meant by "soft" is very broad, just as "e" does not stand only for "electronic," but encompasses the Internet, digital technology, and the future. Soft also includes intangibles, such as value, innovation, product quality, and so on. These are the things that Asia desperately needs.

Discussion

Attaining a balance between innovation and efficiency

Q: **There are many companies in Taiwan with a very rapid pace of development, are not able to innovate effectively because of the pressures to maintain efficiency. How can a balance be achieved between innovation and efficiency?**

A: The only way to do this is to make appropriate adjustments within the organization: to do multi-purpose R&D for different products in order to gain time to innovate. While pursuing innovation, there will be time pressures because the competition in the market is intense and new products must be brought to market quickly. In order to resolve efficiency issues, you need to start from the organization itself. In the process of pursuing innovation, Acer often is faced with the problem of insufficient time or people. However, from another point of view, this is just a challenge that must be met by decision-makers because even if you have enough people within the organization, when sales expand, you'll be short again—and the problems, and pressures, start again.

The employees within an organization will be restricted by time and will be influenced by the duties they have been assigned—is speed the top priority for the particular duties assigned? Or is it the number of patents or sales achieved? As long as the direction of the organization as a whole has been set, everyone can work in this direction. Leaders must change over time and constantly adjust focus—you can't afford to become rigid.

Q: **What is the distinction between internet organizations and non-internet organizations?**

A: Internet organizations and hierarchical organizations are contrasting types, just as the stand-alone computer and the network are contrasting types of computing environments. In a non-internet organization, a subordinate cannot be the master of any task, but must obey the directives of his or her superior. An internet organization is not like this because the subordinate can handle on his or her own 70 or 80% of the tasks. It may even be the case that as long as it's within the rules, a subordinate is free to do anything, and if the rules do not allow some action that would benefit many people, he or she can make a recommendation to people higher up.

As for Acer's organizational structure, a group company's affairs are the domain of that company's CEO and are not passed through the parent company so that the board of directors can authorize the general manager to make decisions. If a problem actually arises, the parent company's influence on the group company is mediated through the board—this is the form of an internet organization. In an internet organization, decision-making is much faster, and the ability to respond flexibly to the market is greatly enhanced.

However, the biggest problem is that it's difficult for a large organization to bring its full capabilities into play because there is no forceful authority to unite everyone and direct them in the same direction. When Acer promoted corporate digitalization, it very quickly made everyone move in the same direction. Unfortunately, if an initiative is directed at a certain customer or line of business, it is extremely difficult to get all operations to put their fullest efforts into coordinating the initiative because each company's priorities are different. Their development focus will be different, and it will be difficult for them to go down the same road.

Handling cross-cultural conflicts

Q: There is a great deal of cultural conflict within multinational companies. Has Acer created any innovative solutions in dealing with such issues?

A: As stated already, innovation must take place on a foundation of rules and knowledge. A business abroad must first conform to local laws and must be adequately informed about local culture. If knowledge is insufficient, there is not enough of a base on which to innovate—and things can't even get started.

Most people believe that the main reason foreign workers are more difficult to manage effectively is an insufficient understanding of them. With insufficient understanding, it's very difficult to convince overseas workers to execute an idea you're set on. Overseas workers are influenced by their local culture and society, and sometimes even the best efforts at persuading them are no match for some outside influence. They often tell me how IBM handles things or what HP or Dell's directions are, and ask why we don't do the same. However, if Acer used U.S. methods to try to beat them at their own game, we'd be doomed to defeat—we need a unique approach. Overseas workers may reluctantly accept my decisions, but have no way to discuss them with me because their background assumptions are different. When I am on site to do the persuading, they may nod their heads, but when I return to my post, because they cannot get a handle on the big picture and can't ask me, it's hard to implement decisions we've made.

Acer's current strategy is very simple—its overseas operations should as much as possible take advantage of Acer's areas of strength and achieve the best possible results for the group, and not try to force things in areas where we don't excel. We don't have to try and do everything at every location because it's a huge challenge to achieve effective management of overseas operations in a short period of time. However, over the long term, when we have a handle on personnel resources and operational capabilities, and the organization is large enough, our overseas operations will naturally have more confidence in us. In other words, if an organization does not have a strategic direction for achieving solidarity with its employees, it is very difficult to build strength.

Propagate know-how through experience-sharing

Q: **If innovation has to be built on a base of knowledge, what is Acer's knowledge management strategy?**

A: We often hear something like this: "Why do the same errors occur again and again? Is there a set of measures that can keep these errors from recurring?" Every company faces the problem of how to transmit experience—such as when the workflow or organizational structure that has been painstakingly built up collapses when experienced people get promoted and new hands take over, leading to the same old errors. In Acer's experience, creating mutual trust and an internal culture free of unwarranted interference makes everyone willing to share their experiences. However, the present era is one in which the volume of information is exploding, and there is too much knowledge to be conveyed verbally, so Acer is already thinking in the direction of digital information infrastructure as it deals with the problem of experience transmission.

In the past, knowledge was limited, and experiences could be conveyed orally. Now, knowledge is everywhere, and oral transmission is no longer workable. Therefore, you cannot rely solely on culture, but must establish a structure through which knowledge can be effectively transmitted and applied. At present, Acer uses experience-sharing at inter-departmental or inter-company meetings to exchange know-how; however, there is as yet no breakthrough method.

Q: **In general, people believe that innovation in management is really just improvement, while only technological breakthroughs represent true innovation. Where is the line between improvement and innovation?**

A: On the distinction between innovation and improvement, my understanding is similar to that of traditional Japanese notions. One of the objectives of innovation is to bring about an improvement, and Japan emphasizes a perpetual quest for ever more refinement—whether it is a car, DRAM, precision machinery, or a hope for something more economical, more sophisticated, or more effective—through a process of constant improvement. U.S. innovation is very different in an interesting way. Its improvements are made in leaps rather than increments—discarding old modes of thought to create a new innovative method. The result may be better than the original, may carry a lower cost, may be quicker, or may even make the original product obsolete and create a whole new picture.

Many people think of technology and management as two axes in a company's operations, but they are together in the sense that technological innovations can provide greater value in the marketplace when supported by managerial innovation. In fact, the importance of managerial innovation is often neglected, even though it often has a greater impact on a company's performance in the market than technological innovation.

Rules are the basis for innovation

Q: **What negative influences can social education have on the ability to innovate? What can individuals and businesses do?**

A: Under a democratic government, alternative thinking is more common, and it slowly helps to create an environment conducive to innovation. Many people feel that young people are difficult to teach, but I don't hold this view. They definitely have more of the qualities necessary for innovative thinking because modern society is more open than before. However, even though modern societies tolerate innovation, order is also an essential element. Order is the foundation on which innovation must be built; without it innovation cannot be channeled in the direction of creating value—it just becomes empty novelty. If individuals and organizations within a society can successfully maintain both order and innovation, then over time their paths will not veer toward either extreme of chaos or stagnation.

Q: **How can employees be encouraged to innovate? What incentives does Acer provide? Are monetary incentives or promotions appropriate measures?**

A: In order to encourage innovation to make the transition to actual intellectual property, meaning patents, Acer was the first company in Taiwan to establish a patent law department, and it had the largest number of staff. In the early stages, this department's chief task was to understand the relevant laws in order to avoid infringing intellectual property rights—this is a concrete example of what is meant by order.

In 1983, Acer was involved in a conflict with Apple Computer, Inc. over intellectual property rights related to the Micro Professor II product and suffered great losses. We made a special effort in the area of intellectual property law and even asked a team from the U.S. to give classes here, teaching us exactly what intellectual property rights are all about.

At the time, in order to encourage the creation of more intellectual property at Acer, we encouraged employees to write reports on other people's patents as a means to disseminate ideas and stimulate innovative thinking. If they did, they were given a reward. There were also rewards for coming up with a patentable idea, for writing up and sending a patent application, and for actually receiving a patent. This set of incentives was quite effective, and we won the National Invention Award for two straight years.

Promoting invention does not only require an appropriate system, but also a means to educate because knowledge is very important to innovation. In the early years, Acer had an additional means of encouraging innovation, which was to have people in the legal department who specialized in writing up patent applications for engineers; however, these are just formal procedures for encouraging innovation.

Room for innovation and the business environment are closely related

However, people have been troubled recently by the fear that PC development has reached a bottleneck and there is little possibility of further improvement. Because of the influence of the larger PC industry environment, Acer Inc.'s thinking in the area of innovation has been greatly reduced. However, Acer Communications & Multimedia (formerly Acer Peripherals) has much more room to show off its capabilities—it is not restricted and can search for new things everywhere. Therefore, we can see that the amount of room for innovation is closely related to the business environment.

Acer has tried to innovate in many different areas, such as improving workflows. The difficulty of this is greater: non-technical problems involve people and not just things—this must be changed. Technological innovation can be moved by a single person, but management is focused on people, so if you are not an executive, and have no power, all you can do to innovate in improving a workflow or changing a thinking mode is to make a suggestion. Unfortunately,

Eliminating the Obstacles to Innovation

other people may ignore this suggestion no matter how wise. Managerial innovation requires an executive, and the higher the person the greater the scope that can be encompassed by innovation. The effectiveness of an organization must depend on the authority of such people.

Q: How can an innovative concept be presented and actualized in a workflow? How can a business evaluate the value and risk associated with an innovation?

A: Anyone who wants to innovate must have confidence and be knowledgeable—otherwise, he or she won't dare to innovate because there will be many challenges to face in subsequent stages, especially in innovations involving people. If the people under you don't submit or take what you say to be idle talk, it will naturally be difficult to institute the innovation. The implementation of all innovations takes place in stages—regardless of how many failures there are along the way, the result must make everyone feel that the organization has a system that can be followed. In the end, innovation must be reflected in action, and if that action is making product, then this process of innovation entails a risk. The reason for encouraging innovation is that it may create great value in the market. As long as this condition is satisfied, much innovation can be undertaken. The fundamental reason why many Asian companies are limited in their innovations is their inability to take on risk.

Innovation must possess an objective and value

Innovation carries risks; however, if there is an approach that minimizes risk it's worth forging ahead. Contrarian thinking is an example of such an approach—think it through without taking action: use time but don't spend money. With a system in place, an innovative concept can be subjected to appropriate evaluation. First, there is order based on rules, then evaluation can take place, not only lowering risk, but also preserving opportunities to develop innovations.

The problem is that once a concept is presented, many people will say "it won't work," and this kills innovation. The reason is very simple: most people will think, "Will it really work?" "If we do that, it will mean a lot of work." "That would be very tiring," and so on. They have not thought on another level: if this innovation is workable, it will not only provide the world with an improvement, but it will also change the way the world functions. Many people have given up before they have even thought about this level; it can be likened to a patient thinking that an operation would be dangerous or exhausting without even thinking that after the operation his or her condition could improve.

Objectives and values are what is needed before an innovation can be accepted. First, there is the innovation's objective, and then thought can be given to possible problems that must be faced, and how they are to be resolved. Second, there is the issue of whether the value of the innovation is greater than the cost of realizing that value. It must be remembered that when thinking about an innovation, future applications must be considered: the experience gained in realizing an innovation is generally even more valuable than the result produced by the innovation—not to mention the fact that experience can be constantly reapplied. If the cost of an innovation is one dollar, and the profit that can be gained from it is fifty cents, but the innovation can be reused 100 times, that profit is actually fifty dollars, making the innovation worth pursuing. When calculating costs, applications must also be considered for there to be an incentive to innovate, and for there to be a willingness to take on risk.

Workflows that can be constantly reused

Q: **How does Acer turn innovative ideas or new discoveries into product specifications and set a time to introduce such a product to the market?**

A: For a great majority of local products, similar versions can be found on the international market, so when there is a new idea or a way is found to lower costs—or alternatively, if the market is felt to be large enough to make a company willing to enter it—you must start thinking about what you will do after you have entered the market. The U.S. has some large companies, such as 3M, that before introducing a new product on to the market first present a product proposal, then form an evaluation committee to decide whether or not to actually produce it. This system has little chance of working in this part of Asia.

Most local companies come to a consensus internally, or follow the directives of the person at the top, or the employees make a recommendation that is supported by the person in command, and then think of a way to complete the task. At Acer, the most important things in developing a new technology are the process of turning it into a marketable product and the maturity of the product design. As a result, we demand a set workflow and, before beginning, set objectives—at what time certain milestones must be reached, and what workflows must be followed in order to reach them—we then make adjustments as needed according to actual circumstances as they develop over time. Though product quality may not have reached the standard targeted, or technical specifications may not have attained the objectives originally set—for example, the product launch time may be delayed—what is important is that we have the ability to progress from an idea to a marketable product. The company can constantly reuse this workflow. Additionally, along with development capabilities, the company must also have mass production and sales capabilities. While developing products, it must also think about how to create a comprehensive workflow, seek out partner vendors, or use a complete application strategy. Each product, or each company, will not necessarily use an identical approach.

Use contrarian thinking to look for reasons

Q: Contrarian thinking helps in producing creative ideas, but the way people think is the result of a gradual accumulation over time; how can established modes of thinking be redirected?

A: To give an example, I don't easily trust others because if I delegate too much to someone who doesn't fulfill his or her responsibilities, I lose out in a big way—this is an example of "straight" thinking. However, thinking of it from the other side, if I want to get more done but am not willing to delegate and invite many other people to work with me, how much can I really accomplish? If I do everything myself, I will exhaust myself; and if I don't want this to happen, then I must trust others. To trust others, I must find the reasons to convince myself that there is no way other than putting my trust in other people.

The risk of trusting others is very great, and to take on this risk means using contrarian thinking as much as possible to find justifications; that is, to make trusting seem reasonable. Reasoning like this takes place in the brain and requires no money. After constantly applying this style of thinking, you can choose the approach with the lowest risk, and after giving that a try and reflecting on the results, you can gradually enlarge the scale. Contrarian thinking requires supporters, so if you cannot persuade everyone to adopt this style of thinking together, you really shouldn't try to go it alone.

Part 3

Globalization Alternatives

CHAPTER 7
An Asian Approach to Globalization

THE APPROACHES EMPLOYED by businesses in globalizing differ according to the nature and role of the company. The globalization experiences of leading multinationals can be useful references, but they should not be imitated. There is no perfect universal globalization principle—Asia must have its own distinctive approach and experiences.

During the past ten years, commonly heard words in economic circles have been "globalization" and "liberalization." The slogans are proclaimed loudly, but there are few people discussing what globalization actually is. In fact, globalization has two levels of meaning. One of these is doing business abroad—which is globalization in a geographical sense. The other level is defined in terms of action—in the sense that an international standard must be reached in whatever a company does.

Why is globalization important? Globalization is tied up with extending the life of a business operation. Before economic conditions were liberalized, in a protectionist environment, no matter how poorly run, a business had room to survive as long as it wasn't any worse than other businesses. But after liberalization, any international company could arrive to compete; hence, local businesses must achieve an international standard in order to be competitive. With the global trend toward liberalization, whether or not a company's operations have attained an international standard becomes crucial. As has been repeatedly stressed, services must be localized. Even if the business is just a small grocery store, if its operations are not up to an international standard when faced with the threat from the world's best chain stores, it will not survive. Many markets in Asia are small and local companies are fighting for markets all over the world, so whether in product quality or business models, they must meet international standards.

Globalization requires certain methods, and companies can consider those used by successful Western and Japanese multinationals. However, for Asia the most important thing is to develop an approach that is uniquely suited to its particular circumstances.

APPROACHES TO GLOBALIZATION

As businesses progress from internationalization to globalization, they will face various situations, adapt to different markets, and meet the challenges brought by different cultures. Even more crucially, from the standpoint of manufacturing and technology, if global disintegration is not brought into consideration, and the product put together is not the best in the world, it will have no way to compete.

During the process of globalization, companies should first establish some fundamentals in their local market—such items as personnel training, risk allocation, and even the speed at which returns can be gained from the market all must be taken into account.

Companies should also develop core competitive strengths in their domestic market if they want to expand globally because a company's domestic market plays a key role during its globalization process. Taking the U.S. as an example, because its market is large, if a U.S. company is able to establish scale in its domestic market, it can develop an effective global management system because it has already trained itself for the task in its home market.

Also, the globally oriented thinking within a company must also incorporate the concept of disintegration. All the specialized talents should not be concentrated in company headquarters. The newest trend among multinational companies is the concept of multiple corporate headquarters; for example, a marketing headquarters may be established in both Asia and the U.S., each setting up shop more or less on its own. The functions of the entire business can be distributed to every corner of the world. At present, Acer uses this approach; for example, headquarters for videoconferencing products is in the U.S. because U.S. technology in this area is the most advanced in the world and has the most core competitive strength.

Depending on the particular function, the globalization approach used will also differ. U.S. multinational companies take the U.S. to be their global center, as well as establishing regional centers in each region. Regional headquarters are an extension of U.S. headquarters, which as much as possible allow regional centers decision-making powers—this is the approach that U.S. multinationals have developed over many years.

THE U.S. APPROACH TO GLOBALIZATION

The U.S. possesses the world's largest market, which is its greatest competitive advantage because U.S. companies can hone their core competencies for global operations—scale, service, workforce, finances, image, product development, and so on—in their own home market. It might be easy to talk about global disintegration, but to achieve the ability to manage an operation in this type of business environment requires a great deal of time. The image of U.S. companies in the U.S. market is virtually the same as its global image, which makes globalization easier for U.S. businesses than for companies from other countries.

All a U.S. company needs to do is establish a good business model and workflows in its home market, and then duplicate it in other regions around the world, where they can also attract the world's best talent. Moreover, with just their home market alone, a U.S. company can establish such great economic scale that even if it doesn't expand into global markets it does not risk its survival. The only real adverse consequence is the loss of face and somewhat lower profits—for U.S. companies, overseas markets are just a little extra revenue. In contrast, if most large Taiwanese and most Asian companies do not go after international markets, they are doomed. This reality is reflected in the fact that many Asian companies specialize in OEM and ODM business, and some of them can even completely neglect the domestic market.

THE EUROPEAN APPROACH TO GLOBALIZATION

Though Europe encompasses many countries, once the Euro begins to circulate widely, European companies will be able to view the entire European market as a single domestic market, its creation making globalization much easier for European businesses.

As for the lessons that Europe can give Asia, it must be noted that Asia and Europe are quite different, and it may be impossible to unify Asian countries into a common market as has been done in Europe. However, the greater China region has the qualities needed to create a single large local market, and therefore good relations between mainland China and Taiwan would be extremely favorable to Taiwan's development because mainland China could then serve as Taiwan's home market. Other Asian countries would also benefit because they would not have to be so wary of political considerations in dealing with either Taiwan or mainland China.

THE JAPANESE APPROACH TO GLOBALIZATION

The approach of many Japanese businesses toward globalization can be compared to older computer systems, with everything under centralized control. It is often the case that in U.S. subsidiaries of Japanese multinationals, Americans appear to be in control, but it's actually the Japanese behind the scenes who are running the show. As business operations expand globally, if all tasks depend on the authority of the central headquarters, problems will develop; the situation can be likened to large host computers being slowly made obsolete by client–server and network architectures. Ten years ago, a factory manager in a major Japanese factory in Malaysia lamented that if the company could do things over again and put more effort into localization of its operations, it might very well be running even better, or more effectively. Clearly, a globalization approach based on centralized control is not as agile as one that gives more autonomy to overseas operations.

The Japanese market provides some advantages to its companies. In Japan, the companies with any visibility are all large conglomerates that are involved in many lines of business. Since there are only these few large companies, each can withstand failures without going under—this is unlike the U.S., where a company that performs poorly ends up being taken over. This has resulted in Japanese companies facing a much stiffer competitive environment in its home market than abroad, making them quite strong in foreign markets. Added to this is their large business scale, which enables them to use strategies of attrition in battling over foreign markets.

However, this strategy of wearing down your rivals by expanding market share at the expense of profits doesn't work in the PC market. In most industries, competition between companies plays out over the long term, but it's rare for a company to have a chance to make a comeback once it has lost out in the marketplace. Just the opposite is true of the PC market, where the winners and losers are decided very quickly, but no victory or defeat is lasting. Japanese PC companies often try to apply to their own businesses the strategies that have worked for Japanese companies in other industries; that is, to slowly wear down rivals by focusing relentlessly on expanding market share. However, in the PC industry, a new generation of products appears so quickly that ground rules for competition are constantly changing, with the result that the typical Japanese approach of squeezing out rivals does not work.

Additionally, the Japanese are deeply involved at the technological level and have the qualities necessary to excel in this area. However, they are relatively weak at marketing high-tech. Japan has many products, such as DRAM, for which it uses a thinking mode that equates the amount of investment required with its "value," while Americans emphasize the value that the market recognizes. From the standpoint of customer value, the U.S. approach is advantageous. However, Japan is the world's second largest market, and for many internationalized businesses the domestic market is still the most critical factor for success, especially in the computer industry. Japanese computer companies rely on their home market for around 70 to 80% of their revenues, so even if their international marketing capabilities are relatively weak, it is not fatal.

A COMPARISON OF GLOBALIZATION APPROACHES

The *World Executive Digest* once described the business models of different countries' international companies this way: "The Japanese approach is like mainframe computing, America's like distributed mini-computing, and Europe's like stand-alone computing. The fourth approach is Acer's—from 'global brand, local touch' evolving into a 'client–server organization,' which is one type of network-like organization." The advantage of a network is that regardless of whether it is one-to-one, one-to-many, many-to-one, or many-to-many, communication is made more convenient; its speed and flexibility are superior as well. The most difficult aspect of a network is the management system, and this is the challenge Acer faced in implementing a client–server organizational structure, and which it will face in developing the internet organization structure in the future.

It can be seen from Table 7.1 (see below) that U.S. companies derive only 50% of their total revenues from overseas business; this is true even for large international companies. Eighty percent of the revenues for European companies come from their home markets, where home market is defined in the broad sense of the European common market. Japanese companies only receive 20% of their revenues from foreign markets. Compare these figures with Acer's, which, including distribution of imported goods, obtains only 10% of revenues from domestic sales. As for Acer-made products, 95% of sales are to foreign markets. From this, we can see that the circumstances faced by Taiwanese companies and those in smaller Asian countries are quite daunting. Even being the market leader in their home market does not mean much since the local market only accounts for a tiny proportion of the world market. The only way for Asian businesses to establish themselves on the global scene is to build their own brands.

From the comparisons shown in Table 7.2, we can see that the globalization approach taken by Japanese companies makes for convenient internal management and allows them to more easily develop globalized, standardized products. The disadvantage of this approach is that it is difficult for them to accommodate local requirements, especially in the area of computer application software. The performance of Japanese companies in the software industry has been unexceptional. The original Sony PlayStation was just a game console with a simple user interface, which made it more a consumer electronics product than a computer; this enabled it to become popular. The PlayStation (PS) II is more powerful than a PC, with Sony's intention of making it the first choice as a home computer in the long run. Other than this, it's difficult to see any area in software where a Japanese company stands out.

Table 7.1 Proportion of domestic and foreign sales for international businesses

	Domestic	**International**
• U.S. companies	50%	50%
• European companies	50%	50% (30% in other European countries)
• Japanese companies	80%	20%
• Acer	10%	90%

© 2000 Aspire Academy. All rights reserved.

Table 7.2 Advantages and disadvantages of different globalization approaches

	Pros	Cons
Japan	• Easy to manage internally • Effectively develops a universal product	• Hard to adapt to local needs • Challenges in software and customization
U.S.	• Well-established principles and practices, easy to expand	• No feeling of ownership for overseas units
Europe	• Have strong units in Europe because of autonomy	• Overheads are too high

© 2000 Aspire Academy. All rights reserved.

The advantage of the U.S. approach is that aggressive expansion is allowed in those large markets with more comprehensive management principles in place. Although regional centers have autonomy, if there is a conflict in principles with headquarters, the view of headquarters prevails. Since U.S. companies are so domineering, foreign operations don't have a sense of belonging; in any case, what headquarters says goes. This fails to give anyone else a great sense of accomplishment.

The European approach allows companies to do well in each individual country, but without achieving large economies of scale, preventing them from gaining the benefits of a disintegrated operation model internally. The result is often high overheads. The reason for Philips' recent closure of several plants in Taiwan lies here.

By comparison, Acer's approach to globalization is more effective in accommodating the trend toward global disintegration, achieving speed and flexibility, controlling costs, and even proving effective in business localization and in providing employees with a sense of belonging. The problem is that Taiwan is not the center of the world, making it difficult for operational models that we're most familiar with to become the basis for a global strategy. Moreover, to implement a client–server or internet organizational structure, there is a crucial factor to be considered: the quantity and quality of skilled workers. In order to build an internet organization, we must have many people with a similar level of skills and information management capabilities. When we wanted to extend our client–server operational model to our overseas companies, we discovered that their operational capabilities were insufficient.

CHALLENGES OF GLOBALIZATION FOR ASIAN COMPANIES

It is difficult for most Asian companies to internationalize. Their local markets are likely to be too small, making them unsuitable as a training ground for doing business globally, and it is very difficult to build up a global brand starting from the local market. In addition, there is a shortage of internationalized workers, and there are no similar markets that can be expanded into.

Taking Acer as an example of an Asian multinational, it has around 34,000 employees (including foreign operations), with more than half of these overseas. From this, we can see that the degree of Acer's internationalization is not less than that of a typical U.S. multinational company. In the area of revenue distribution, Taiwan accounts for 10%, Europe for 25%, the U.S. for 40%, and other areas for 25%. In South-east Asia, Acer is number one, but in Asia as a whole its ranking is not high. The reason is that the individual markets in South-east Asia are too small. If a Japanese company is number one in the Japanese market, it is Asia's number one. All that a mainland Chinese company has to do is be number one in mainland China and it surpasses us. If Samsung only sells to the Korean market, it is bigger than us.

Looking at Acer's experience in overseas markets, appearances can be deceiving. Although the U.S. was at one time Acer's largest market for own-brand products, it was also the place where we suffered the greatest financial losses, and a decline in revenues eventually led Acer to exit the consumer PC market there. On the other hand, our share of the notebook computer market in Italy is 31%, and this result was difficult to achieve. The varying performance of the same product in different countries can be attributed to localization—marketing and services must be tailored to the local environment, and in Italy we have a very strong local team.

To summarize, a global operation encompasses the following four areas: manufacturing, research and development (R&D), sales and marketing, and joint ventures. In the past, attention may have focused on just one or two of these, but with the extremely intense competition in today's global markets, a more comprehensive approach is called for, and an effective globalization strategy should take into consideration how to make the most of international resources in all of these key areas.

The first phase of Acer's globalization

On Acer's path to internationalization, the first foreign operation to be set up was a purchasing team, then sales, and next R&D, manufacturing, and so forth. When this was established in 1976, the extent of our international business was buying from foreign vendors. When Acer set up operations in the Hsinchu Science-Based Industrial Park in 1981 and started foreign sales, it was exporting own-brand products. Microtek set up operations in the Hsinchu Science-Based Industrial Park at the same time and also had its own brand. In 1982, we participated in an electronics trade show in Tokyo and exhibited Micro Professor I. A German magazine reported that Micro Professor I was Microtek's product because its name was similar to our name at the time, Multi Tech. Even though it seems funny now, it made me determined to change the name of our brand.

Thanks to innovation, Micro Professor I had an absolute competitive advantage. It became a hit around the world, giving us a great deal of encouragement. Acer's PCs have been market leaders internationally because we took advantage of the momentum created by Micro Professor I and began to internationalize. During this process, we had our share of frustrations, such as when a Singaporean customer told us that Taiwan was not a computer-producing country, so they weren't interested in importing our product (at that time people in Asia thought computers should come from the U.S. or Japan). The acceptance of Acer products in Europe and the U.S. was simpler; in these places only the product itself mattered, and not the country where it was manufactured.

There is an iron rule in running a business: the most effective approaches are the ones that can be repeatedly used. In running Acer, we encountered a problem: each time we tried to expand our market, it was as if we were doing it for the first time. A successful experience in a certain market couldn't be duplicated in another market. The experience in developing a certain product was not applicable in developing another product. This created a certain amount of waste. At the same time, we were expanding and we had to constantly establish new core competencies; additionally, because we were becoming more and more successful, we were led by the market, becoming more and more deeply entrenched, with more and more personnel and material resources being invested. We set up many branch offices and distribution warehouses and gradually started doing marketing locally.

The second phase of Acer's globalization

In 1987, we changed the name of our brand as well as our corporate logo. In 1991, Acer had several problems in its operations and undertook re-engineering. In 1996, we launched an executive development program and trained many top-level managers. When we introduced our new corporate logo in 1987, we loudly proclaimed a new slogan: "Aiming for world number one." At the time, Acer operated on a very small scale, but though our people were few in number, morale was high. Now that the company is large, it's as if morale is no longer at the same heights. When the company's shares went public in 1988, things started to go downhill, and we had no choice but to re-engineer. For a company to re-engineer itself, a monumental task is at hand, and it generally requires many years, often without any assurance of success. In the 1990s, IBM went into decline, and the person charged with managing its re-engineering, CEO Louis Gerstner, spent more than five years on the job before IBM finally began showing some of its old bloom. After DEC's business began to deteriorate, it never made it back. In the early 1990s, Compaq started to decline, but then undertook huge price reductions to regain the position as the world's number one PC vendor. It has started to falter again in the last one or two years, and we don't know what the future will hold.

While Acer was in the midst of its executive development initiative, it occurred to me that after the company expanded, annual growth might slow from 20% to 15%. At the time, I felt that moderating growth rates after the scale of a business had become very large was the norm. However, there were several events that completely changed my mind on this. The growth rates that Compaq and Dell achieved during their peak, as well as Acer after its re-engineering, were in the range of 30 to 50%, even reaching as high as 70%. This made it evident that the growth rate and size of a company were not strongly related, with the growth rates completely dependent on the level of competitiveness. When running a business, if your growth rates cannot exceed the industry average by a large margin, even if you do have positive growth, you're still in decline.

The third phase of Acer's globalization

During the second phase of Acer's internationalization, Acer came up with many new ideas in facing various challenges, comprising what the *World Executive Digest* called the "fourth approach." This worked well at first, but after 1996 problems started to crop up. After 1997, I started to rethink about how to strengthen the brand. First, we began total brand management (TBM), putting it on an equal footing with total quality management (TQM). We stressed that just like product quality, brand management was everyone's responsibility. Beginning with me, every person in the company had to get involved in TBM. Without effective brand management, it is very difficult to establish brand value.

Moreover, the speed of price reductions for PCs is very rapid, as is the rate of product obsolescence; how to operate globally is an issue that requires thorough consideration. When Acer was running its "global brand, local touch" and "client–server organization" internal campaigns, each regional operation had a great deal of autonomy, and there were only some general principles that people needed to consider when dealing with the brand; the actual actions taken were the responsibility of the local operation.

Since 1997, though this way of doing things had its advantages, there were also bottlenecks such as high levels of inventory. Each regional operation needed enough inventory management specialists, and I had to approach this problem from a global viewpoint. In addition, to compete in the PC industry based on product alone is not enough, so customer service, customer-centric operations, and customer relations management (CRM) become very important. The next task was to build on our base at the time and to strengthen intellectual property and service businesses—this created a new approach.

Next came an approach for enabling a client–server organization to deal with adjustments in task allocation in the future. The result of this thinking was the concept of the internet organization. Even though there are many problems in managing this form of organization, its effectiveness is definitely superior to traditional hierarchical organizations.

FROM REGIONAL BUSINESS UNITS TO GLOBAL BUSINESS UNITS

In 1989, when Acer began its internationalization phase, it established two types of operations: strategic business units (SBU) and regional business units (RBU). SBUs took charge of R&D and manufacturing, while RBUs oversaw regional marketing. At the time, globalized companies such as IBM and Hewlett-Packard emphasized the need for businesses to localize, so we used the RBU approach to satisfy this requirement.

Later, however, large multinationals gave up on this approach, instead adopting an approach based on global business units (GBU). Acer's circumstances were even more dire because its SBUs and RBUs were in many cases independent companies. The RBUs and SBUs of U.S. companies were different profit centers within a single company, and when controversy arose over pricing for transactions between different profit centers, headquarters could intervene and resolve the issue immediately. When Acer's SBUs and RBUs squabbled over price, there was no big boss who could step in: each operation had its own boss. As Acer Group chairman I could have intervened, but that would have been unfair—whose side was I on?

Another problem was that the region covered by each operation was not large enough to be integrated as part of a single global strategy; in competing with European or U.S. companies, we couldn't match up. For example, Acer's market share in Malaysia was number one, but if Hewlett-Packard (HP) had resolved to take the number one position, it wouldn't have been too difficult. For HP, Malaysia is just one segment of a global market because in any case HP's operations are based on its overall performance around the world. Therefore, within this global framework, HP would use special resources or pricing to win market share in Malaysia. That was not the case for Acer: each of our regional operations were autonomous and had to rely on themselves to create profits, with the result that it was difficult for Acer to implement a global strategy.

As another example, Acer is the market leader in Latin America, but while large U.S. companies can "dump" unsold inventory in Latin America, Acer does not have this option. The excess inventory accumulated by each regional operation must be dealt with by that operation, and so again, no global strategy can take shape. Add to this the increasing complexity that industry competition is giving rise to, and if a regional operation is not competitive and it cannot sustain itself, then SBUs and RBUs do not have the ability to remain independent businesses. Besides, everyone was originally part of one company but had become SBUs and RBUs, each with its own ideas and self-interests. In order to change global operational workflows, the amount of time needed for negotiations would be enormous. Therefore, Acer came up with its "21 in 21" (twenty-one publicly listed companies in the twenty-first century) initiative. This idea was arrived at within the framework of the client–server organization concept. At the time, many of the operations that Acer planned to make public were overseas RBUs, though currently there are fewer opportunities to take foreign subsidiaries public. On the other hand, the fact that the number of publicly listed companies in the Acer Group has not decreased is due to our adoption of the GBU approach. Any independently operating company possesses a global business scope; it can do its own marketing or use overseas operations that were established previously; contracting a competitor or a third party is also feasible when thinking about global strategy from the standpoint of products.

At present, many multinational companies are using this type of approach, with the head of each of their regional operations directly responsible for handling public relations. Taking HP as an example, the general manager of its Taiwan regional office isn't in charge of products or sales strategies; he's just there "watching the house," handling public statements and coordinating internal communications, without needing to worry about operational results. Each of HP's product lines has its own head product manager who reports directly to headquarters in the U.S. Right now, IBM also uses this approach.

In shifting from SBUs and RBUs to GBUs, much resistance will be encountered. To gain some insight into how to handle the process, we even visited the regional headquarters of a foreign multinational in Singapore and saw how difficult the shift would be.

GBUs use product operations as the basis for conceptualizing their business structure. To resolve issues concerning operational expenses, they use regional operations (RO) in place of RBUs. Therefore, the general manager of a multinational's foreign branch is the head of national operations (NO) for that particular country and not the person in charge of product operations. In this manner, there are various product teams under each RO and NO, all integrated with a GBU; this approach is what's known as matrix management.

Matrix management is the most difficult approach for most Asians—Westerners tend to be more disciplined and are very clear about who's responsible for what and who's playing what role; in contrast, the roles that Asians play socially or in the workplace are often very ambiguous or even conflicting. The people lower in the organization report to the GBU in terms of business, and to the NO from the point of view of localized management issues, such as information management, accounting, finance, and other administrative functions. Product competition uses an end-to-end thinking approach. Competition in the past was just a matter of doing the product itself well, but that's no longer the case. A product's competitiveness, operational efficiency, market strategies, and where markets are segmented all influence the final success or failure of a product. Therefore, everything must be managed by the GBU itself, which takes responsibility for success or failure.

CONCLUSIONS

The Japanese and U.S. experiences can serve as reference points for Asian companies, but should not be imitated. If an Asian company merely copies their approach, it can only aspire to being a second- or third-rank company at best. Asia must have its own unique approach, and this is the reason for Acer's spirit of aggressively undertaking to develop new approaches.

On the road to globalization, most of Asia faces relatively few problems in the areas of manufacturing and R&D, while its challenges in marketing are the greatest. For Asian companies to internationalize, a large number of internationalized talents are needed. However, Asian companies should not feel too discouraged by this: they are not the only one facing this challenge. The businesses in every country have their own burdens, but there is no choice but to try and overcome them by constantly making adjustments and searching for new approaches.

It is most important to have the desire to be a global citizen: during the process of globalization, wherever you go you become a good corporate citizen. Positioning yourself as a global citizen, and maintaining an attitude of thinking of international operations from the point of view of how they can benefit the local country, is very important.

DISCUSSION

Q: During the process of internationalizing Acer, how were you able to systematically nurture talent?

A: When Acer trains people, it wants them to reach an international standard, with a sufficient understanding of products and technology. I have to admit that what Acer relies on is actual work experience and not a very strong codified training program.

In early 1980, Acer sent some people abroad, and these people were pleased with the opportunity to live and work in a foreign country, especially for as long as two years. At the time, I believed that handling international tasks should be done on a rotating basis.

Later, I discovered that because during these two years the staffing requirements in our domestic companies had changed, there were not necessarily any suitable positions for all the people coming back—and we lost many good people as a result. Many of the international talents in Taiwan's computer industry are employees that Acer lost during that period. In response, Acer revised its system of assigning employees to work abroad, changing the terms to three to five years. After this change was instituted, the largest source of problems became the employees' children; after receiving such a large part of their education abroad, they couldn't fit in well when they returned home. As far removed from typical business considerations as this problem may seem, it will most likely be a major challenge to Asia in general as it internationalizes.

Talented workers are the key to internationalization

In addition to actual experience, Acer did attempt to provide formal educational training in internationalization. Acer's training center considered different training courses, but their success was very limited. The only thing that seemed to be effective was actually assigning people to overseas posts. Acer currently has offices in more than thirty countries, but many of them don't appeal to people. We even tried to provide extra incentives to encourage people to accept postings to such countries, but the results were less than ideal. To resolve this problem, investment is required; first of all, our business requires internationalized workers; and second, in the long term a company must train new talent. In any case, where this talent ends up going is not important. Even if they return home and do some other type of work, their experience will still help in more effectively using the local resources. From this viewpoint, business should actively invest in personnel training over the long term, without immediate regard for how effective it might prove to be, because worker talent is a key for Asian companies' future internationalization.

Additionally, Asian internationalization strategies also need some further refinements. In the area of manufacturing, Taiwan's internationalization challenges are minimal. When companies have gone to mainland China, South-east Asia, or Mexico, the results have been quite good because manufacturing can be effectively controlled. However, in internationalizing manufacturing operations, we relied on many Malaysians. Taiwanese workers are willing to be posted to the U.S. or other advanced nations, but to convince them to leave home for other places is not as easy. Therefore, we had to rely on Malaysian internationalized workers. For Taiwanese companies to build their own brands and make a mark in the U.S. or Europe, they need to develop some basic strengths—they should first do the job in mainland China and South-east Asia.

Running a business well is the means of giving back to the local country

Q: If a company wants to be a good global citizen, the most important task is to give something back to the local country. Does Acer have an effective approach?

A: I have always felt that running the company well means making a contribution back to the local country. Not to make improper profits, not to exploit workers, but to improve living conditions and the economic competitiveness of the local area—these are all ways of contributing back. For example, when Taiwanese companies go to mainland China and provide training to their local workers, it has a real effect on the workers' lives and produces a very large influence on the local area. "Giving back" does not necessarily mean making charitable donations to the locals; it's just as Acer does in Taiwan itself: the best way we can contribute to Taiwan is by running Acer well, training workers, being a global citizen, and doing our job well. In running a business, your starting point should be how to be a global citizen, and not making the earning of money your only priority; only in this way can you avoid exploiting workers, destroying the environment, and avoiding your responsibilities as an institution within a society.

Q: What are some approaches to internationalization that Asian businesses can use?

A: In terms of function, internationalization can apply to manufacturing, R&D, marketing, and so on. However, workers cannot be classified using such a simple scheme. For example, people working in finance must not only understand Asia's financial environment, but also understand the local banks and financial relationships. Or take administrative workers—they must understand local laws, the labor market, how to recruit people, how to train them, and so on. This type of experience requires international operations in order to gain a thorough knowledge. From a manufacturing and R&D point of view, the various regions of the world don't differ that much; however, marketing is another matter. To internationalize marketing means understanding markets, and each region's market is different.

Considerations in using acquisitions to attain internationalization

Q: By acquiring U.S. companies, Asian businesses can acquire a brand, marketing capabilities, and so forth, as a way of becoming internationalized. Are there any issues that need particular attention?

A: Before considering an acquisition, a company must test the following assumptions. First, that it will be able to actually obtain the acquired company's brand, its marketing avenues, and personnel talent. Second, that the products acquired are competitive and able to fit in with a globalized product strategy.

Even if both assumptions are fulfilled, there are still several problems that may arise. First, there is the issue of how marketing plans can gain a consensus from product people since product strategies are extremely broad. If you can obtain a consensus, it'll be easy to focus efforts; but if marketing and product people have their own ideas, it will create many problems. Even more, sales and marketing people tend to be more optimistic in their sales forecasts, but in this industry an overestimate means having to deal with excess inventory, and inventory management is critical.

The second issue that can arise is if there is a shortage of a product that is selling well and a surplus of another product that can't be moved. How to promote sales of the slow-moving product is also very important because if sales people continue to push the product that's selling well, you still end up losing money. A third issue is that sales management costs may be too high. In general, U.S. companies first create a sales plan and then do things according to this plan. Even if business conditions change, they will not stop; resources used in a market where such companies predominate are less effectively used.

Coordinating brand management

Q: During the internationalization process, how can a company's executives and employees communicate effectively?

A: Communication is necessary for any task, so front-line people and the back office must have dialogue—this is the reason we turned our larger operations into GBUs. As mentioned previously, "global" has two different meanings. One is internationalization in a geographical sense, and the second is using a comprehensive approach from the beginning to the end of each task. A GBU's responsibility extends from the beginning to the end, and it is solely responsible for its success or failure.

Acer's PC and scanner operations are both GBUs, and the communication between them is very limited. They must take charge of their own work, so networked management is greatly simplified. In the end, the only part that must be coordinated is the brand management issue. Using GBUs, the chief source of conflict arises from global brand management—how is this operation defined, compared to another? Pricing will cause conflict, and this requires coordination of brand management. Other areas, such as sharing of management experiences, are not necessary for survival.

To be an international business, you must make each business a complete, independent unit, and the only problems that may become divisive are those that arise from everyone using the same Acer brand.

Q: Domestic markets for many Asian companies are too small, so what do you believe is the ideal internationalization approach for them?

A: Every internationalization approach has its pluses and minuses, but because circumstances in different countries are different, the internationalization approaches of the U.S., Japan, Europe, or other countries may not necessarily be suitable, so companies must develop their own approaches. Although the Acer Group's GBUs do not necessarily employ the same internationalization approach, we will at least develop the same starting point for thinking about issues of common concern. These starting points can serve as the principles for Asian business's internationalization.

The first principle is that even though to internationalize a brand in the short term you don't have to follow the European or U.S.

approach too closely, you must create a strong brand image in Western markets. Use Western industry magazines to introduce your products, participate in their product competitions, and win awards, but do not actually sell in Europe or the U.S.—make them yearn to buy your products but be unable to buy them. The more you sell to the European or the U.S. markets, the more money you lose because after the sale your responsibilities begin—so making Europe or the U.S. unable to buy your products is not necessarily a bad thing.

You can also do ODM business or work with a foreign company to jointly develop a brand. This is not an easy thing to do. In 1986, in order to break into the U.S. market, I was ready to seek out some well-known U.S. and Japanese companies to jointly develop a brand. I had discovered that Japanese companies couldn't break into the market, and while some U.S. companies had a good brand, they didn't have the technology or products, and so they couldn't get into the market either. I sought them out for cooperation, to see if there was any chance of creating a dual brand. Even if I didn't earn any money from the deal, it wouldn't matter as long as I could gain a brand image. The local partner could take any profits, and I could use this brand to win shares in the world's other markets.

However, up to this point there's no solution for creating an effective computer brand in the U.S., unless your products are really "different." And this something "different" must have real value if it's going to be of any use. Unfortunately, the PC is just what it is—there's no room for being different.

No way to duplicate the "Italian experience"

Q: **Acer has a very strong team in Italy. Why can't the approach used there be applied in Germany, England, and other European countries?**

A: During the earlier stages, Acer did very well in Europe. Northern European countries are small, so we could let our local partners do the legwork, and they obtained good results for us. However, no matter how good these markets are, they're limited. Later, Acer's European office was in Germany, so we also did quite well in Germany, with our notebook computers reaching number three.

As for Italy, in the past we consistently did very poorly, and we had no choice but to change course. We switched agents and were sued by our original agent. The reason why the outlook in Italy completely changed is very simple—it was because when we bought Texas Instruments' notebook division, we inherited their team in Italy. Because our products were much stronger than Texas Instruments', this team became stronger and stronger; and because of the effectiveness of this team, the product line was also expanded.

This approach can be duplicated in other markets, but only by redeploying the leaders. You can't transplant the employees below them; and without a very good team, it's very difficult to fight a battle over a market.

Q: How do you make elements in the internationalization concept you advocate, such as TBM, known to all employees worldwide?

A: In the early stages, Acer did a good job in this area. The company's internal publications and communications all promoted Acer events and ideas, and this was quite effective. As we expanded geographically, language barrier and time lapses resulted. Added to this was the now enormous size of the organization—you couldn't expect that one sentence from higher up would penetrate to everyone lower down in the organization: what was involved was not just a problem of words, but the actions that went with them. This was the biggest challenge.

Therefore, when we instituted TBM, we started from scratch—the brand vision, mission, and personality all had to be drawn up from scratch, and then education and training had to be provided using various tools. In the future, when we implement company-wide initiatives, we will adopt something similar to the approach used at Cisco. Their CEO John Chambers takes advantage of video-conferencing technology to make his directives viewable over the company's internal networks. This enables the company to implement changes extremely quickly. The problem is that to communicate the spirit of an idea, the language you use is critical. If we use English, the spirit will be compromised: there are too many things that even you yourself are still trying to formulate, and you can't describe it clearly using Chinese, to say nothing of English.

CHAPTER 8

Acer's Breakthrough in the International Arena

WHEN SURVEYING CURRENT TRENDS in global economic development, it is evident that running a top-tier company not only requires seeing the whole world as a market, but also making the strength of your credentials for competing globally an important index of performance. The ultimate objective of globalization is to effectively exploit worldwide industry disintegration and international resources to enhance business competitiveness. Therefore, globalization considerations and implementations are increasingly important to the running of a business. Put simply, this is the condition that must be met to achieve comparative advantage and competitiveness. In the process of evaluating themselves, businesses should not only consider short-term results; direct, indirect, and long-term factors also must be brought into consideration.

On the subject of internationalization, the most controversial viewpoint is the so-called "industrial hollowing-out." This viewpoint became established in the U.S. in the 1980s when a shrinking of the manufacturing sector led to worries that the country's industrial infrastructure was being weakened; later, the Japanese discovered that their own industry was shifting manufacturing overseas, with the result that Japan itself would become "hollowed out."

However, if you look at the internationalization approach in the U.S., industry shifts operations overseas while, domestically, companies still look to work with subcontractors; the more operations that are shifted overseas, the more competitive they become. Whether the hollowing-out idea really makes sense depends on whom you ask. There are many factors that need to be considered as a company internationalizes, but the ultimate objective of globalization is to effectively exploit worldwide industry disintegration and international resources to enhance business competitiveness.

Reasons for globalization

The direct motivation for companies to undertake globalization typically relates to overly high labor costs in their home country, a shortage of workers, and the need to take advantage of materials or natural resources overseas. Additionally, globalization is a road that must be taken in order to get closer to markets, overcome protectionism, and save on logistical costs. In summary, to return to the competitiveness formula we discussed earlier, globalization is needed to expand markets and lower costs. The most basic starting point for globalization is the creation of value and the lowering of costs, and in fact this is just the basic law of supply and demand—manufacturing is the supply, and the market is the demand. Business globalization reflects the relationship between the two.

Location considerations

When businesses go overseas for development, the most important consideration is location. When choosing a location, people tend to think only about the local physical infrastructure. Physical infrastructure issues are becoming easier and easier to resolve thanks to the current scrutiny of the situation in different countries. All that is needed is a plan, along with a willingness to spend money, and within a few years, construction of transportation and other public facilities can be completed. However, "soft" infrastructure, such as education, government policy, industrial structure, and so on, all require ten, twenty, or even more years to complete. During the process of globalization, soft infrastructure is more important than physical infrastructure. Technical workers and human expertise are also very important types of soft infrastructure.

There is another factor to scrutinize: whether the region offers incentive measures, with incentives being of two basic types. The first type allows the retention of more profits for investment or to take back to your home country, and the second type means assistance required by your operations, such as personnel training. At the same time, political factors, labor union activities, and the aggregation of related companies must be considered also.

It is worth looking at the situation in Europe. Most European countries have a socialist tendency and are most concerned with the level of employment. In addition, European trade unions are very powerful, and labor costs are very high, with the result that unless the purpose is to establish a new market, Europe is not a good place for a low-cost manufacturing operation. However, in order to attract investors, European countries often offer more assistance than Asian countries. For example, when some local companies established factories in Europe, they benefited from extremely preferential treatment, so much so that they were able to set up their manufacturing sites without spending virtually any money at all, while the local government even took responsibility for a portion of personnel training—it looked almost as if one couldn't ask for any better treatment. However, it was a trap. This is because Asia does not control much of the market, and most of the Asian businesses that set up shop in Europe are primarily manufacturers. The local governments were thinking from the point of view of raising employment levels, so they would provide assistance proportionate to the number of jobs the company could create. The problem was that in order to operate over the long term, the more workers a company hired, the heavier its burden—if for any reason sales were not up to expectations, the companies were in big trouble.

Taiwan also provides investment incentives, but it uses a "scholarship" style of incentive—European incentives are subsidies. The "scholarship" style of incentive means that only businesses that perform well can receive assistance. The European subsidies, on the other hand, are given regardless of a company's performance. It doesn't matter that the more people you hire, the heavier your losses become—the government still provides subsidies because you are helping to solve its employment problem. Europe's approach has caused some companies to become dependent on subsidies, and this system is not necessarily appropriate. As usual, when economic decisions are dominated by political considerations, the results are less than ideal.

Manufacturing site considerations

When choosing a manufacturing site, it is important to stress that low labor and land costs are not the most critical factors and should not be overemphasized. The most important thing is that the location be convenient logistically, that it offers high-quality engineers and general workers, and that it has a related industrial aggregation nearby. In addition, as mentioned, government incentive measures and political factors must also be considered.

In 1990, Acer began preparing the establishment of manufacturing operations overseas. Before this, Taiwan's small- and medium-size businesses had already moved much manufacturing to South-east Asia or mainland China because they could not tolerate labor shortages and high costs in Taiwan. When Acer began thinking of moving manufacturing offshore, we immediately considered Malaysia's Penang: the U.S. and Japanese electronics industries had already cultivated the area for more than twenty years; not only were the engineers of high quality, but skills needed for factory automation even surpassed those in Taiwan. In addition, the area was gradually developing its own suppliers, and the Penang government was predominantly Chinese, providing us with favorable tax breaks and other incentives. When Acer set up in Penang, it was already toward the tail end of the great exodus of Taiwanese manufacturing. We knew that Malaysia was gradually developing labor shortages, but still decided to go ahead: we believed that the added value of Acer products was high, and that labor shortages would only affect other companies, and not us.

In 1995, Acer set up in Subic Bay in the Philippines, and I was a member of the first team to go there to survey conditions. I have a special relationship with the Philippines as I am the chairman of the Asia Management Academy, which is located there. I go to Manila almost every year to conduct meetings and have close contacts with many key businesspeople there. Previously, when I interacted with these businesspeople, I was left with the impression that they lacked confidence in the Philippines. After Ramos took office as president, the investment climate improved, and on one visit I found that the local businesspeople were already beginning to invest locally. If they were willing, then so was I. Therefore I took a look at Subic Bay—which has a good infrastructure and is close to Taipei—and quickly made the decision. Later events showed that Subic Bay is in fact a good place for investment. This anecdote implies that intangibles that build confidence are also a very important basis for consideration. When selecting a manufacturing site, you cannot only look for the cheapest place.

I was not personally involved in Acer's decision to establish a factory in Suzhou in mainland China. However, after comprehending the factors that went into making the choice, I supported the decision. The reason is Acer did not want to crowd into southern China along with many other Taiwanese vendors. In addition, while in terms of future sales within mainland China, Shanghai can serve as a base to cover the entire market, in Shanghai, Acer was not considered very important: all the world's big companies are there. Further, labor is expensive, and it's not easy to obtain land. We therefore selected Suzhou as the site of our factory. There is an aggregation of electronics companies near Suzhou, and in terms of components-sourcing, transportation, and worker quality, Suzhou is an excellent site—add to this the high quality of life in Suzhou, which is the only way to attract people from Taiwan to go there. Later, Acer also set up a site in Chungshan in Guangdong province, but not in Dongwan or Shenzhen because these two places are overdeveloped. The main factor behind our choice of Chungshan for a factory is convenience for foreign sales, thanks to its proximity to air cargo facilities in Hong Kong and Macao. In addition, Chungshan is a famous scenic city in mainland China, with a high standard of living.

When Acer established a factory in Mexico, the main consideration was proximity to the U.S. market. Acer used a twin-cities approach for its manufacturing sites, with one site in El Paso, Texas, and the other in Juarez, Mexico. There is also another twin-cities site, in Mexicali and across the border in California. The idea behind a twin-cities site was that high-level executives could live in the U.S. while the actual workplace was in Mexico, with workers hired in Mexico.

R&D CENTER CONSIDERATIONS

The main consideration for research and development is the quantity and quality of talent. An R&D site must also be able to integrate with local technology centers; for example, semiconductor design in Silicon Valley and software in Seattle. In telecommunications, Israel is very strong. As far as R&D is concerned, a company can't go wrong by going to where the skilled workers are. High-tech companies are competing for worker talent, comparing whose engineers have the greater expertise. Therefore, in the future, companies must think of ways to take better advantage of mainland China and India, which represent enormous untapped resources of potential R&D talent. In choosing among locations that meet the basic condition of having sufficient quality and quantity of skilled workers, the main considerations are cost and whether or not there is a culture and environment that encourages innovation.

Acer has several well-known products that we developed before IBM did, such as a 32-bit computer, which was developed by Taiwanese teams sent to Silicon Valley. The Acer Aspire was also developed in the U.S., an innovative product designed to meet market needs. The decisive factor that enabled the introduction of these products was not related to people but to location. Recently, Acer established an NT$250 million venture capital company, and it will invest half of these funds in the U.S., which is the world's center for high-tech R&D. Acer has also set up a software center in Shanghai. My greatest hope is that in the future, thousands of engineers working in Acer's Aspire Park here in Taiwan will be doing world-class technology development.

Marketing site considerations

Technology and manufacturing need to take into account globalized operational considerations, but marketing support must be localized because when an appropriate market scale is reached, localized marketing operations become essential. Once a company has several marketing centers operating, regional headquarters is needed, but when exactly should one be set up? How large should it be? Should it be partly or fully autonomous? Should its legal status be a representative office, branch office, subsidiary, or joint venture? These questions demand careful thought.

When Acer first considered setting up operations in Europe, the Japanese had already established several sites in Dusseldorf, Germany, and large Taiwanese companies were also there, so we went along with them. Following the crowd has many advantages; in addition to the location already having been subjected to careful scrutiny by others, industry aggregations were already well established, and any difficulties could be handled through mutual assistance. However, Acer's European headquarters once moved from Germany to the Netherlands, with the main reason being that in Germany, language and culture gave rise to some problems that made life for people sent overseas from Taiwan very inconvenient. As for choosing Silicon Valley for our regional headquarters, the environment is almost identical to Taiwan, making things very simple. We selected Miami for the sake of the Latin American market as at least half of Miami's population speaks Spanish. In the

Middle East, we chose Dubai for our regional headquarters because it is the most internationalized country in the area.

In choosing a location for a marketing office, all the following issues must be considered carefully: capabilities in sales promotion and inventory, distribution, customer service, technical support, and perhaps extending to simple assembly, product development, or even to whether the location can serve as a regional shipping center.

Operational Headquarters Considerations

When choosing a location for operational headquarters, the personnel factor must be carefully weighed because the headquarters needs to be capable of many different functions. For example, whether or not the site has advertising firms that can provide adequate localized marketing and advertising must be considered. From this perspective, Singapore is superior to Taiwan; Taiwan's advertising companies can handle the greater China area, but Singapore's can provide a professional level of services for English-speaking markets as well. For headquarters, whether a location has the needed personnel resources is the most important factor.

As far as government incentive measures are concerned, Singapore offers an OHQ (operational headquarters) incentives package, with five- or ten-year tax exemptions. There are even incentive measures for manufacturing operations outside of Singapore, as long as the required portion of Singaporeans are hired.

For employees sent abroad, the most important things are the local living conditions and work environment. Whether cargo, people, and capital can move conveniently in and out is also important. In view of this, with the exception of the high-tech industry where Taiwan's personnel resources and industry structure are superior to Hong Kong's and Singapore's, at least for now, using Taiwan for the regional headquarters is exceedingly problematic.

International Capital Considerations

Global capital moves rapidly, and it makes a big difference whether a company's liquidity is based on capital or debt. The Asian financial crisis, especially in South-east Asia and Korea, was the result of international expansion funded by debt. Taking on debt has the advantage of low cost and avoidance of the necessity to share profits with a bank, but while it allows a company to retain control, it can lead to instability, and just a little bad luck can lead to a loss of its financial base.

On the other hand, by introducing international capital, the organizations providing capital become, over the long term, shareholders in the company. This involvement enhances the invested company's image as being internationalized and benefits its reputation in its home country because it has demonstrated that recognized international institutions are willing to make it a target for investment. Operating this way saves costs: when there are no profits being made, banks still require payments; but if it's an international capital investment and no profits are being made, the money stays in the company, and there isn't much negative impact.

ACER'S EXPERIENCE WITH GLOBALIZATION OF CAPITAL

Using Acer as an example, before going public in 1987, I had already sought foreign capital investments, including Citibank, Chase Manhattan, Sumimoto Bank, H&Q, China Development Corporation (CDC), Prudential, and so on. At the time, this was a new way of doing things, and it won trust in the company, giving the public optimism about Acer's stock. Reflecting on it now, my only regret was that aside from CDC and Sumimoto Bank, the investors made their money and sold out. This was different from what I'd expected: I'd been looking for long-term investors. At the time, I lacked experience and didn't know H&Q's and Citibank's money came from venture capitalists, whose strategy is to get out when they've made their profits. Since then, when I look for investors, I've been especially careful to find people who are in it for the long-term.

Later, Acer again introduced foreign capital into the company: one source was convertible bonds and the other was global depository receipts (GDR). Though company debt must be paid off eventually, it can be converted and at the right time it can become capital. GDR is capital right from the start. Up to now, all the convertible bonds Acer has created is in the form of capital. This approach has the advantage to the company that for the same amount of money coming in, the amount of stock issued is less. To do this requires a lot of legal documentation, and two or three months before the stock issue the preparatory work has to begin.

Recently, Acer received an extremely favorable response when it issued convertible company bonds overseas. The company had already prepared all the necessary materials and was ready to go to cities abroad and give road shows to promote the bonds, but before we started, we found out that US$300 million of the convertible bonds had already been sold, without even having to present any briefings.

Two years ago, Acer had a GDR issue that created a new approach. The original idea behind GDRs is to raise funds from outside, but with half the GDRs being bought by the company itself—something like hoarding its own stock—which then become stock options for employees. Because Taiwan does not have stock options, I used this approach to provide incentives for the executive teams at our overseas operations. Our stocks in Taiwan are used to share dividends with employees in Taiwan, while for employees overseas we use GDR options as an incentive—all of which is perfectly legal.

To raise the capital needed for global operations, regardless of whether for sales or manufacturing, requires the cooperation of local financial institutions. However, in tight local markets it is not a simple matter to establish an independent line of credit, unless the parent company makes a total guarantee, or the parent company's own line of credit can be transferred. Acer had a very unique experience in Malaysia. Malaysia's central bank has some funds available for export industries, but to protect its domestic banks, Malaysia does not allow foreign banks to get into this business. Acer did not feel it needed to put that much capital into its Malaysian operations, so it contracted Citibank to create a type of special stock issue. Even though a special stock issue is a way of borrowing money, it is also a form of capital stock; and the larger your capital stock, the more money you can borrow from Malaysia's central bank. Because these preferred shares could be bought back, they were also like cash, and therefore this foreign bank could do business. This was a new approach and was created within the legal framework.

In 1992 and 1993, Acer came up with its "21 in 21" plan, whose goal was to have twenty-one public companies in the twenty-first century. At the time, the feeling was that it would be possible to go public in many countries, but later we discovered that even though it was easy to go public overseas, in practice results were less than ideal. While going public is beneficial for enhancing corporate image, it has a negative impact on the price-to-earnings ratio, and therefore it upsets employees. Employees work equally hard, and their levels of achievements are similar, but in different places they are compensated differently—giving rise to feelings of bitterness. Add to this the very strict regulations governing conflicts of interest, and when a public company makes policy decisions, any person who might have a financial stake must avoid any questionable involvement—leading to many flaws in the management process. So we decided to get delisted. Overseas, it's easy to go public but difficult to go private again. In Singapore, to protect the interests of investors, the government requires that at least six months be spent in getting delisted, and this doesn't include the preliminary procedures. Approval must also be obtained at a shareholders' meeting, the courts must grant permission, and so on.

The portion of the "21 in 21" plan applicable to Taiwan was very successful, but worked less than ideally in Singapore. In 1996, I was originally going to push Acer to go public in the U.S., but because our sales there were not very strong, we took the GBU approach instead. At present, GBUs with Taiwan as an operations center can very quickly go public, so it makes the goal of twenty-one publicly listed companies a realizable target.

ACER'S EXPERIENCE WITH GLOBAL MANUFACTURING OPERATIONS

Acer's experience with global manufacturing has been very favorable. At the very beginning, we sent local employees to staff an overseas site, but at our first overseas manufacturing site in Malaysia there are now only a few of these people left; the rest are local people, and we've transferred the needed technology and know-how to the local managers. In fact, we are already taking advantage of Malaysia's own engineers to set up operations in Mexico, mainland China's Suchow, and other places. However, when we wanted to take advantage of local talents in Subic Bay in the Philippines, a problem arose: it is difficult for Filipinos to gain permission to go to the U.S., especially for women (the fear is that once they get to the U.S., they won't be willing to go back to the Philippines). Apart from this, Filipino talent meets expected standards; whether in the area of quality, productivity, or cost, they are very competitive.

Taiwan has world-class credentials in manufacturing and as a supplier; and because in the past ten years Taiwanese companies have effectively globalized their operations, they have also built up a large customer base. As an example of this process at work, it's worth mentioning the case of Acer Communications & Multimedia (ACM, formerly Acer Peripherals). The people that ACM sent abroad returned to Taiwan after a few years:

after gaining valuable experience overseas, they could take on positions high up. This is different from the situation in the past, when we would send marketing people to Europe, but because they didn't really succeed there wasn't a place for them after returning to Taiwan. However, people involved in manufacturing who were sent abroad developed very nicely. At present, the general managers at Acer Display Technology, Darfon, and Acer Media Technology were all people sent to Malaysia who returned to Taiwan.

ACER'S EXPERIENCE WITH INTERNATIONAL MARKETING

International marketing in advanced countries is extremely arduous, while in developing nations it's relatively simple. The main reason is that in advanced countries the biggest challenge is overcoming the customer service problem—it's extremely difficult to attract top talent, and the dilemma of who listens to whom crops up. Very self-proud nationals may not think they need to listen to the views of people from elsewhere, and this will give rise to a lack of trust between headquarters and the particular overseas operation. This problem does not occur only in the PC industry, but in other segments of the electronics industry.

To address this problem, I developed a "peripheral attack" strategy. Acer's PCs and systems had been successful in smaller markets, and after 1995 we believed it was time to go on the offensive against the central market—meaning the U.S. In the end we failed, the reason being that our resources and capabilities were inadequate for the task, especially considering that we were fighting for a geographically distant market, which immediately put us at a disadvantage. However, Acer will utilize the "peripheral attack" strategy again, but this time the term *peripheral* will not designate only a geographical location but also product lines. According to this concept, PCs are central, while peripheral devices and components make up the periphery. Acer's central product (PCs) could not win the central markets of the U.S. and Japan, but this time Acer will use "peripheral" products to attack these markets. Right now, we are using the basic brand established in the U.S. with our PCs to sell computer peripherals quite successfully.

This new "peripheral attack" approach can also be applied in Japan and Korea. Although Korea is not a core market, the ethnocentrism of the Koreans makes it extremely difficult for foreign vendors to enter their market. Japan makes very high demands on product quality, after-sales service, and distributors, so when we decided to enter the Japanese market, I was prepared to spend eight years doing battle. Unfortunately, after eight years we still failed. However, in the last two years Acer has used the "peripheral attack" strategy to sell components in Japan and has succeeded in making a profit. The "peripheral attack" is a very effective approach, but big companies must win central markets; and unless they do, they have little control of their own destinies.

Using central products to pursue central markets is an extremely difficult and thankless job; it requires an enormous commitment of resources, with heavy investment in advertising and after-sales service, as well as the need for a very large distribution network. However, when doing "peripheral" products like components and computer peripherals, you can take advantage of the brand created for central products, and there are fewer competitors—you can thus take market share very easily. If a company uses only a small fraction of its total powers, it can take advantage of "peripheral" products to win in the biggest markets. This again points to the fact that in running a business, companies shouldn't try to force things, but work from their strengths.

Acer's experience with globalized R&D

The U.S. possesses an environment for innovation, and this is reflected by the fact that the Acer Aspire PC was developed there. Even though the U.S. is very innovative, the process of commercializing an innovation requires a division-of-labor approach with Asian companies. The commercialization process is very complicated, and you need a very good procedure and sufficient experience.

As for business models applied to the commercialization process, ODM has proved to be superior to OEM. OEM means a U.S. company, for example, completes a design and passes it to an Asian company for manufacturing. For ODM, the Asian company does the design, with the finished product then sold under the U.S. company's brand. If all the U.S. company does is design, and the manufacturing is done in Asia, this type of OEM approach will lose to an ODM approach. The reason is that commercialization is not simply a matter of manufacturing alone; it involves end-to-end control, from design and materials to product quality and costs. These things are not decided by manufacturing, but should already be decided at the design stage.

In a pure OEM model, when a U.S. company completes a design it doesn't know the manufacturing conditions; it's not familiar with suppliers of raw materials and components—this results in higher costs as compared to those in an ODM approach. To pin down the essential difference between OEM and ODM, one could say that the former is more segmented, while the latter is holistic.

Even though ODM still can't match a start-to-finish OBM (own-brand manufacturing) approach, the fact remains that the possession of ODM capabilities provides a powerful advantage. Without such capabilities, Taiwan's industry would long ago have been surpassed by Malaysia.

It is extremely difficult to reconcile the views of the overseas R&D operations and headquarters. SBU executives must take charge of commercialization, which includes planning, resource allocation, and prioritization—so the overseas R&D center cannot be allowed to work on its own, but must accept the authority of headquarters.

Conclusions

Deciding on a location for an overseas manufacturing operation is not as simple as choosing a location for a sales office. This is especially true when manufacturing personnel have to be transferred overseas; hence, an appropriate scope for the operation must be carefully considered. After selecting a location, managing the operation is relatively easy: the workflow, regulations, and management approach are similar almost everywhere.

However, Asian marketing capabilities have not reached a high standard, and this can be problematic if Asian companies want to establish more than just a manufacturing presence in Western countries. If they hire top local talent, it's impossible to manage them effectively because they'll want to go their own way. Naturally, settling for second-rate talent defeats the purpose of trying to tap local workers to establish strong marketing capabilities. To manage effectively an operation in a Western country is not only a problem of different languages, but it also involves different cultural values and ways of thinking—the result is that it's very difficult for Asian companies to get good results.

In choosing a location for overseas operations, labor costs are not the key factor. Industry structure is more important than the cost of labor. The most difficult aspect of Taiwan's internationalization has been its shortage of marketing talent, and this is an area in which we must invest more aggressively. Cultivating marketing talent costs more money, requires more time, and carries a higher risk than training manufacturing talent. The U.S. business school ranked number one in international management, the Thunderbird Business School, was established in 1964 when many businesses in the U.S. wanted to go international. They discovered that local professional talent for the effort was lacking, so the school was founded. From this example, we can see that training professional talent is the key to every internationalization.

Discussion

Q: **Acer has established a software research center in Shanghai. How were preliminary evaluations carried out? How was its performance after establishment assessed?**

A: Taking advantage of mainland China's software engineering talent pool is a long-term plan. In terms of expertise, Taiwan definitely doesn't have sufficient resources, which is why we thought of going to China to train people there. At first, the only places we could consider were Beijing and Shanghai. We felt that software engineers should also have a market sense, so we chose Shanghai because it was closer to core markets. In the early stages, we sought software engineers with an eye on the international market, but from a long-term perspective we will more and more need engineers who understand the local business development and market requirements—and Shanghai met these needs as well. Additionally, Shanghai has many universities, and there is a lot of professional talent available. From the time of establishment to the present, two years have passed, and we feel positive about the results. We hope that after establishing a base there, we can expand into many other locations.

Effective management is the biggest problem

For Asian vendors establishing foreign operations, capital is not the main problem. The biggest problem is effective management. The question is: should you send a large number of your own people over to manage the operation there, or should you train managerial talent from the local population? This is an issue that requires careful consideration. In addition, the biggest question in considering whether or not to cooperate directly with local software companies is whether or not intellectual property rights will be respected. This takes a very long time to establish and implement.

Acer has a set of software strategies that are quite different from what we have used in our manufacturing businesses; many of these strategies rely on venture capital investments. In the past few years, besides making our established manufacturing operations the focus of investment, we have also extended investment to the development of intellectual property and the development of service businesses (especially Web services)—this strategy is already set. The key to developing intellectual property is not in having many people, but in having strong capabilities; therefore, we have used venture capital investment to get in early, with the amount invested equal to around 15 to 20% of the total invested in a company, and not just 3%. Acer is using real involvement to develop intellectual property, and in the future its software business will use this approach. Additionally, because software development is so dependent on the commitment and enthusiasm of the people involved, we have agreed to allow employees to hold 50% or more of a company's stock.

Q: In the area of intellectual property, what methods does the company have for preventing employees from gaining expertise and then leaving the company?

A: From the standpoint of accumulating experience and skills alone, people are free to do as they please. A person can freely choose to go to any company he or she wants; this is just a basic human right. We merely use the common international practice, which is employment contracts, to stipulate that intellectual property produced at the company becomes the company's property. As for the actual process, any ideas or experiences that a person accumulates and that are not covered under the legal definition of intellectual property belong to him or her personally; and there is nothing to be concerned about here. Looking at it from another perspective, if someone you've trained is able to contribute to society, even if it is through another company, you should feel a sense of achievement.

Constantly create new opportunities for people to demonstrate their capabilities

Q: After a business internationalizes, how do you think about promotional opportunities for employees returning from overseas assignments?

A: The key is that you have to constantly expand opportunities at the parent company. If there are no opportunities, and a person wants to leave, from the standpoint of management it may be best to let him or her do so. With too many people doing too little, when everyone gets together they just end up getting in each other's way. The Acer Group's most distinctive feature is that it is constantly creating opportunities for people to demonstrate their capabilities. Four people who returned to Taiwan from assignments in Malaysia found positions as general manager or chief financial officer in Acer companies. If a business cannot continue to develop and just stays on its original territory, with so many people crowded into a static space, there is no room for cooperation; the only solution is to develop new opportunities within the organization.

Q: **How many publicly listed companies does Acer now have? Which ones have been more successful? How can conflicts of interest be avoided?**
A: Acer has six companies in Taiwan that have gone public, and two overseas. The Singapore company has already been delisted. We already have 89% of the stock in the Mexico subsidiary, and after buying the remaining 11% it will also be delisted.

As for which companies have been most successful, Acer Sertek was originally the parent company for the entire Acer Group, but it later became a subsidiary of Acer Incorporated. ACM was originally the parent company of Acer Display Technologies, but during this year the latter's capital base will probably surpass ACM's. There is no question of good or bad in this, just as each generation in a family stands on its own merits; each company sustains its operations, and all have development potential and should constantly create new startups—short-term results cannot be used to judge success or failure.

Resolving conflicts of interest between group companies is very simple: all the companies involved together hold less than half the seats on the group's board of directors, which has the final say. The board of directors uses majority rule, and representatives of the parent company are kept to a minority by principle. If during the decision-making process there is an obvious conflict of interest, the involved parties immediately withdraw from the discussion. Obeying this principle avoids conflicts of interest between related companies.

Avoid battles that you cannot afford to lose

Q: When choosing a site for an overseas operation, how can you assess the local investment risk?

A: Twenty years ago, I said repeatedly not to fight any battles that you can't afford to lose. Putting aside for the moment the question of internationalization, even in your home country investment and expansion carry risks. When considering investment risks, you have to adhere to the principle that even if the investment is a total failure, it won't destroy the whole company. I use this same principle in my personal investments. When I became involved in Acer, I had already thought through this and reassured myself that even if the whole venture was a failure, I wouldn't lose everything.

When Acer began investing overseas, as in Malaysia, we already had enough scale in Taiwan that if the overseas venture was a bust, it would just mean a slowdown in our expansion; even if we suffered losses, it wouldn't affect the big picture. When Acer moved all its monitor production out of Taiwan, it already had operations in mainland China and Malaysia. If one day Acer moves all its manufacturing overseas, we will definitely consider making the quantity and quality of operations in Taiwan higher than those overseas: this is where our center is; if our Taiwan operations permanently lose manufacturing functions, it won't have any negative impact. This principle is illustrated by the movement of U.S. manufacturing to Asia. At this point, there's no turning back; but the U.S. companies don't suffer from this because they have taken their operations to a higher level, concentrating on brand marketing, value-added services, and so forth. Our desire to maintain a secure foundation explains why Acer moved into mainland China later than everyone else; it was only after we had very solid operations in Malaysia and the Philippines that we could feel comfortable about investing in China, and at that point we didn't have to worry so much about political turmoil.

Concerning technology R&D, it's a process of very gradual accumulation. If some key personnel leave, the point of reference is still at the company—software, patents, and documents still remain—and this allows people lower down to take up where the departed people left off. This is the basis of the company's core competitiveness. The technological competitiveness needed to maintain a company's business has to be in its own hands.

Q: How can multinational companies insert themselves into local business/government networks?

A: Acer's relationship with the government in Taiwan is very normal, and the same is true of Acer's overseas operations. A relationship is established based on the scope of our investment and the contribution it makes to the local society. We think about all our investments from the perspective of being a global citizen and providing benefits to the local area. We don't go out of our way to try to nurture a special relationship with the local government—in the long term, doing things that way just doesn't work. By following these principles, Acer has always been warmly received wherever it has gone.

First, train your people in your home market

Q: Taiwan's marketing capabilities are third rate, but how can Acer be number one in Asia under these circumstances?

A: Acer's marketing capabilities within Asia can't be called third rate: even the U.S. companies with the best marketing capabilities don't do well when they get to Asia. That's because they don't understand the Asian markets. They may be able to hire top executives, but when the team as a whole does marketing in Asia, they can't beat us. In hardware, Acer aims to be the world number one—that is our ultimate goal. When I talk about Acer services winning Asian markets, the final objective is also to be number one in Asia.

Manufacturing PCs is a very tough business, and we haven't been able to be number one in Japan, Korea, or mainland China. But outside of these countries, Acer has been number one in Asia for quite some time. If in the future, Acer can be among the top few vendors in China, or even number one, there is still a chance to be number one in Asia as a whole. You must make this kind of preparation, have this kind of ambition. As for when we might reach the goal, because there are factors beyond our control, we still need to make efforts over the long term. However, even if today the opportunity already existed, we would still not be fully prepared since we haven't achieved the necessary scale. To be number one, you must train your people in your home market, and when the time comes, you can see just how far you've progressed.

Acer has the potential to be number one in many areas, whether in terms of geographical location or product categories. U.S. companies don't have the ability to be involved in as many areas as Acer, while the competitiveness of Japanese and Korean conglomerates is compromised by not being made of many ferociously competitive smaller companies as Acer is. The internet organization structure that is taking shape at Acer is not only the right approach for Acer and Asia, but also for other smaller nations. The strategy of using small size to win big is one we must continue to develop. However, if there is only Acer fighting it out in the international arena, our strength is not enough; if more companies join in, our strength will be greater.

CHAPTER 9
From OEM to OBM

FROM THE PERSPECTIVE of long-term business models and future development, OEM and OBM can in fact exist together. Businesses must place importance on OEM, just as they must do for OBM. Both are business opportunities, and they share a large common base of know-how.

The biggest challenge for Asian businesses and companies in other smaller countries as they internationalize is brand management. If their local markets were large enough, such as mainland China or the U.S., creating a strong brand would be easier. Creating the Acer brand required us to pay the most dues and was the biggest challenge we've faced. Looking ahead, Acer must continue to invest in the brand.

To establish an international brand, the first priority is to gain a large market share in countries nearby. The next thing on the list is to diversify the product lineup. Taking Acer as an example, we not only manufactured PCs, but also had a battle plan based on a broad range of products. In this area, the problems that businesses with small home markets will confront have already been met and overcome.

OEM VERSUS OBM

It's very easy for Asian companies to do OEM. All that's needed is to establish appropriate capabilities and to work cheaper than somebody else, and the OEM customers will come to you. If you add in better design, you're even more competitive. The obstacles to entering the OEM business are relatively few, and you don't need as many core competencies. What's more, OEM customers are very large, so a company that does OEM can very quickly establish scale. After entering the information technology industry, labor costs had risen, so the source of Taiwanese companies' competitiveness has shifted from manufacturing to design. We have established this new competitiveness through inexpensive engineering talent, rapid design capabilities, and a greater willingness to show flexibility in accommodating OEM customers' specific needs. Thanks to our ability to keep costs relatively low, there are still reasonable profits to be made through OEM.

The problems with OEM become evident over the long term. Maintaining strong performance over a long period is difficult: reliance on specific customers is very high; most OEM companies have only two or three customers, and it may even be the case that a single customer provides more than half of total revenues. This situation gives rise to a very high level of psychological pressure: should an order be cancelled you will suffer heavily. In the early stages, where business was won primarily by providing the lowest cost, the competition for OEM business was very intense. In order to obtain OEM customers, Asian companies often got involved in irregular business relationships. In today's information technology industry, this problem has already become much less common.

The reason that OEM customers have looked so favorably on Asian companies is that we are very specialized and very effective. However, if we want to diversify and get into other lines of business or do new products, it upsets our OEM customers. For their OEM vendor to have its own brand is even more difficult for OEM customers to accept. Good examples are Taiwan's Giant and Kennex: they first did OEM and then tried to establish their own brands, but making the transition was not easy. Acer was just the opposite; we first created our own brand, and once we had done that there was really nothing an OEM customer could do about it.

In general, the major transitions to be made in moving from OEM to OBM (own-brand manufacturing) can be summarized as: from simple to complex management, from large scale to small, from short time frames to long, from few customers to many. OBM is a much more complicated business than OEM, and management is correspondingly more difficult.

Whereas the scale in OEM business is large, OBM means establishing a customer base of one customer at a time, meaning that it is not easy to achieve a large scale. Shifting from OEM to OBM means shifting from short time frames to long: when a company is doing OEM, all that needs to happen is for the customer to place an order, and you can do the production and deliver; but with OBM there are many preliminaries, after-sales service, and other steps to take care of, so the time frame is lengthened. Customers also change from being few in number to many. Doing OEM, you may only have two or three customers, and it's easy to provide service; but with OBM you may well have tens, or thousands, or even tens of thousands of customers, and it may be impossible to deal with all of them. Therefore, in making the transition from OEM to OBM, the challenges are formidable. However, even though the number of OEM customers is few, the customers have many different choices, and it's very easy for them to make changes in policy; the result is that your relationship with customers is not close. When doing OEM business, there is also no way to create intangible assets such as those that a strong brand represents.

THE CHALLENGES OF OBM

For companies with small home markets, to build their own brands is a great challenge. Taking Taiwanese companies as an example, their home market is not only too small, but their products and technology are not innovative enough. These factors make it more difficult to establish a strong image in the broader marketplace. If they use price-cutting as a battle plan, it's even tougher. Therefore, Taiwan has few companies with experience in building an international brand.

In the past, Taiwan's image, as expressed in reactions to the "Made in Taiwan" label, was poor, and while the situation is much better now, the view among consumers has not changed very much. In computer-related components, such as motherboards, regardless of whether it's quality, design, service, or flexibility, Taiwan has the best reputation with other computer companies around the world. However, computers pass the local OEM and are put together by system integrators elsewhere, so consumers have no way of understanding the excellent reputation of the Taiwanese products that went into the making of their computer. From this perspective, Taiwanese companies have it easier: they deal with experts and with a smaller number of people. If you are dealing with non-experts, you must have greater powers of persuasion; and to deal with a large public, you must spend a great deal in advertising fees. Taiwanese companies do not have the capabilities to communicate directly with consumers at this point.

Taiwanese companies also lack long-term successful precedents as they try to build their own brands. Up to the present, the only examples of relatively well-known companies that have performed reasonably well with their own brands are Acer and Giant bicycles, while others, like Proton, Kennex, and Travel Fox, all failed in the U.S. market. More than ten years ago, the Own-Brand Vendors' Association was established by these five companies to actively promote branding, and in the early stages it could be considered successful. However, we all hit the wall in the U.S. at around the same time. The U.S. market is really a difficult market to crack, and the risks are high because the competition is very intense. Taiwan cannot depend on a single successful brand to build up the whole country's image—unlike Germany, where if one company makes great cars, the image of German cars in general is enhanced.

The OEM business is a completely different story. Each segment of the OEM industry has been a success—notebooks, monitors, and motherboards—and Taiwan can probably accommodate five companies in each category. However, OEM customers will not all choose the same supplier and will naturally help build up second and third companies doing OEM in the same segment. If there is one company doing OEM successfully in one segment, OEM customers will want there to be a second and third choice, so it's easy for Taiwan to do OEM. Even in principle, establishing a leading brand in the market is not easy, so I sought out several OEM suppliers as partners to fight the battle cooperatively.

Aside from these factors, the resources available to Taiwanese companies are limited, and it's very difficult to beat out competitors over the long term in fighting a war of attrition like building a brand. However, when a new industry arises, it's easier to establish a brand, with Yahoo! being a good example.

Strategies for OBM companies

If a company wants to establish its own brand, it must innovate and create actual value through this innovation. U.S. products are very innovative and expensive; even when only a minority of people see their value while most do not, the building of a brand can begin, and companies can still survive. The U.S. has many companies that can survive at this phase, but when their product becomes a mass market item, they can't keep it up—unless they work with an Asian OEM partner.

Building your own brand requires long-range planning and a step-by-step approach. If you apply the "peripheral attack" strategy, described in Chapter 8, in building a brand, the first step is to build a brand in a region with smaller, less competitive markets where fewer resources are required. The second step is to produce a niche product, such as the Micro Professor learning tool that Acer used early on to establish its brand. However, from a long-term perspective, it's Acer's PC brand that built up momentum for its brands of monitor, scanner, and CD-ROM drive.

If resources are limited, you can build up a brand through a niche product or geographical market, with the condition that you must have enough wins in different niches because success in a niche does not carry the prestige and significance of success in a major market, nor does it carry the benefits of image and confidence of the latter. Moreover, success in one niche does not imply success in another—being number one in Taiwan does not mean you'll be number one in Hong Kong. However, if a company can be number one in a major product category such as PC systems or in the U.S. market, the effect can carry over into other product or geographical areas. The mainland China market represents another springboard that Asian companies, and particularly Taiwanese, can take advantage of in building a brand with global impact.

With the current scale of Taiwanese companies and others of similar size, trying to establish a brand in the U.S. is not necessarily advisable: doing business in the U.S. is very risky, and sales overheads are very high. However, companies can take a more casual approach. Taking the Taiwanese company Proton as an example, it concentrated on the local market but went to the U.S. to build up its image and gain publicity by winning an award, getting some good coverage in the industry press—all without having to commit much in terms of personnel or financial resources.

In going after a market when using limited resources, finding a local partner is a very important consideration. The problem is: why should they work with you? They must be able to earn a profit by selling your product; and the precondition for a product being profitable is that it be innovative and have market appeal, with high margins and a competitive price; or they simply have to have confidence in you.

Brand strength must be accumulated over a long period of time, especially since a strong brand image is like possessing something visual—when you see the brand and you close your eyes, you still have an impression of it. If your eyes are closed too long though, the image disappears. A brand must be constantly repeated at appropriate times, and the greatest current difficulty is that brands are seen but don't make an impression. There are too many brands on the market, and unless a brand has some outstanding feature or is extremely innovative, giving people a visceral shock, like Sony's Aibo robot dog or Acer's Aspire PC, it won't be noticed amid all the different brands out there. Therefore, products and technology with innovalue become ever more important.

Entering a price war is the biggest taboo in building a brand. Regardless of the industry, a company cannot succeed over the long term using cutthroat pricing as a strategy, unless it is able to redefine customer expectations simultaneously, as Southwest Airlines has done—providing low-cost fares, but also lowering customer expectations for the extent of services they will receive. Otherwise, trying to maintain profits by reducing manufacturing or material costs is fruitless: this is a capability that any competitive company possesses, and starting a price war just means everyone loses.

In the market, a brand is an invisible asset, and it is constantly providing an intangible sense of value to consumers. Having an established brand does not imply the need to provide extra value to consumers, but rather to lower costs while maintaining a reasonable level of value. A strong brand allows unit costs to decline steeply because it gives the company a more powerful bargaining position in doing business with other companies, and it provides its own advertising effect on consumers, thereby saving on marketing costs. A strong brand also gives you advantages such as making it easier to get loans or to hire the best talent—and these coincidentally also lower costs.

When physical objects are mass produced, unit costs are also somewhat lowered, but there is a big risk in doing this: if you are not careful, you may overproduce. There has never been an exact equilibrium between supply and demand in the market, and hence imbalance is normal. Therefore, mass production of physical goods implies risk. If after investing in a physical plant, initial costs cannot be recovered before the depreciation period expires, manufacturing capacity is left idle, and this is a burden that must be carried into the future. Building invisible assets also carries risk, but it's carried by the investor and influences the stock price. The burden created by investment in a physical plant is heavier than for invisible assets; you can get out at any time after investing in invisible assets, but when you buy an equipment that will depreciate in value there's no going back.

For a company, its brand is a treasure. In early 1990, Acer started losing money, and foreign banks withdrew funding. However, domestic banks all continued to support us, and we can thank the Acer brand for this. The foreign press boldly guessed that it was the government propping us up behind the scenes, as Acer was Taiwan's only well-known brand, but this was completely groundless. From this we can see just how valuable a brand is: it's an asset you can rely on when your very survival is at stake.

There are many theories stating that only by investing well in advertising and establishing a brand can products command a high price. This idea is a trap and will increase future spending: when a brand image does not have that value yet, but products are priced too high, consumers willing to buy will become fewer and fewer. I'm very wary of the idea that once you have a brand you can start charging high prices, and I have also received a painful lesson in this area. In March 2000, Acer won a big victory in the mainland China market: the number of desktop PCs we sold was five or six times higher than in the past. Formerly, Acer had not been able to crack the mainland China market because Acer thought of itself as a world brand and wanted to go up against IBM, Compaq, and HP, and our prices were higher than those of China's local vendor Legend. Positioning ourselves so high left us with no room to lower costs. Later, we started using a strategy of "international brand, local pricing," using the Acer brand but selling at Legend's prices, and we discovered that not only did margins not decrease, but they actually rose. This is a matter of business attitude; if you think having a brand means you have to set high prices, you're sure to lose and have no hope of sustaining your business over the long term.

Brand-naming considerations

A brand has to make an impression on people, and it's best if it's simple and easy to remember. The current thinking about brand-naming is not to create a separate logo, but to combine the name and logo into one. Earlier thinking called for a logo; however, IBM and Dell have no separate logo, but just typographical effects with the letters in their names. The

IBM name has endured for a long time and has become synonymous with computers—every consumer recognizes it. In mainland China, everyone recognizes these three English letters, and to them IBM has already become a Chinese word. The impact of brand-naming is in the name itself, so as much as possible a simple English name should be used. The best-known Japanese international brand is Sony; on the other hand, the names of two Japanese multinationals, Matsushita and Mitsubishi, sound almost alike. Acer's Chinese and English names were arrived at separately—Chinese is Chinese, and the English does not have to be a translation of its meaning or transliteration of its pronunciation.

Also, when building an international brand, registering a trademark is very important; both the brand name and the trademark itself should be registered. When we changed our English name to Acer in 1987, we registered the name in more than 100 countries. Everything went smoothly except for England, where there was an architectural firm also called Acer. I want especially to emphasize that a brand name and a corporate identity system can be constantly changed. I haven't seen any company that has never changed its name or trademark. The reason is very simple: when a company is created, no thought is given to the day when the company becomes "great"—so there's no harm in changing the name. Some companies insist on not changing their name—many Japanese companies are like this—this is a traditional way of thinking and is not appropriate for present times. The reason for changing a company's name might be that the company has already become very different, that its business is no longer the same, or that it wants to welcome a greater future. What's more, the company's original image may not be favorable; to create a new image, a change in company name or trademark is worth considering.

In April 2000, Acer held the e-Life Show to demonstrate how a PC company had thoroughly transformed itself into a digital corporation and also re-engineered its image into one with greater advantage in the future. This was important because, except for Dell standing as the standard example of successful marketing innovation, PC companies are associated with a lack of innovation, as being just a bunch of "me-too" companies. Computer companies have to constantly think about their image, otherwise there will be a very negative impact on future sales.

Branding is a deep and broad subject. Companies have to consider whether to use a single brand name to cover a number of products, as IBM and Sony do, or to use multiple brand names, like Procter &

Gamble. Using multiple brands is especially well-suited for trend-oriented products; after a certain product's image becomes popular, and you are thinking of applying this brand to a new product, it's best to use a different brand: the pricing and market segmentation are different. The trendiest cosmetic brand in Taiwan these days is SK-II, but hardly anyone knows which company produces it. In any case, that company is one with multiple brands. However, automobile brands don't work like this. Car companies use the sub-brand approach. After one brand has been established, a series of sub-brands is introduced. Managing a brand requires great care, otherwise confusion may arise. Even if the company itself doesn't sense what is happening, it may have a negative impact on the consumer market, which will affect future business.

Two years ago, the Acer Group introduced total brand management (TBM) (see Figure 9.1), and the reason we decided to redefine the brand was that we had not effectively and consistently conveyed our image to the general public. We did some deep thinking about what exactly our objectives were, because if you're not clear on the brand image you want to create, anything you do may conflict with that image and all the effort is wasted. Acer has been in business for more than twenty years, but looking ahead to the coming decades, or even centuries, we had to ask, what is the mission that Acer is seeking to complete? We found that the

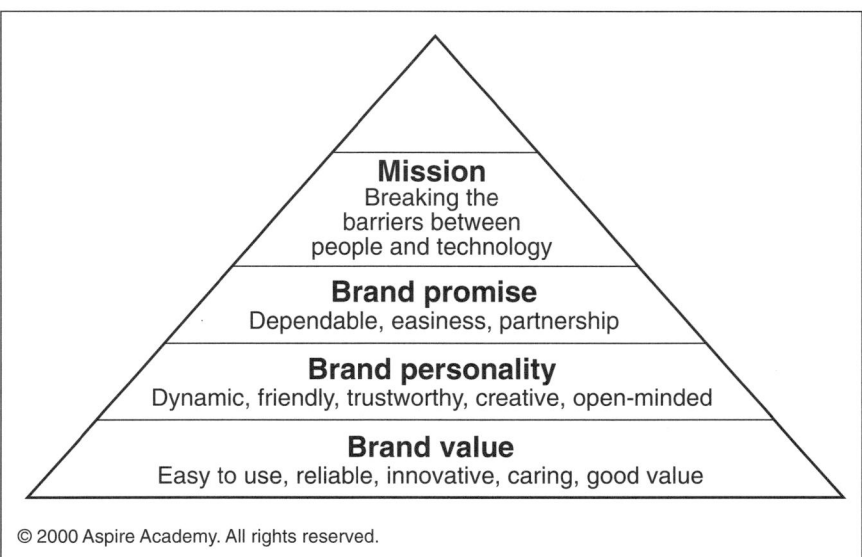

Figure 9.1 Acer's total brand management (1)

fundamental reason for Acer's existence is to allow everyone to enjoy fresh technology and eliminate the obstacles between people and technology. Our mission is not just to develop technology, but to pay attention to customer needs; whether creating new products or services, these obstacles must be eliminated.

In addition, what promises does Acer's brand make to consumers? We have three types of customers: first, OEM, meaning other companies in the industry, and our promise to them is to be a reliable partner; second, corporations, and our promise to them is that we are and will be a company they can depend on; and third, individual consumers, for whom we represent convenience and simplicity.

As for Acer's brand personality, we drew up a list of ten or twenty distinctive features, and from these chose five: dynamic, friendly, trustworthy, extremely creative, and very open. Finally, we asked ourselves: for consumers, what is the value of the Acer brand? The value that Acer provides is ease of use, dependability, innovation, concern for users, and good value for the money. Every company should be able to provide these types of value, and every company has its own distinctive character. Each type of product at different times has a different aspect that must be highlighted—in the past, the ease of use of high-tech products was often stressed, and more recently, styling has become a focus—and adjustments have to be made in response to changes in the larger environment to enable you to concentrate your efforts in the right areas.

Acer's TBM

TBM is a new type of approach, and it is emerging in the same way as early quality control efforts, which eventually led to the development of total quality control. Brand image is in fact a very complicated concept, and establishing a brand is one of a company's core functions. In the early stages, Acer's efforts to build a brand were piecemeal and relied totally on inspiration; however, doing things this way didn't allow people to work together as a team, so the company had to have a planned, systematic approach to establishing a brand image in the market. When we work on a brand strategy, we always go to different countries to make assessments, whether this is done through market surveys or focus groups. Adjustments are then made for each market. According to the surveys carried out by Acer, Acer's brand efforts have made it number one in Taiwan, number two in South-east Asia, and number three in Europe. But in the U.S. very few people have heard of Acer. What we're after in total brand management is results in the market.

After several years of effort in managing the brand, the question that came up in the end was: what exactly does Acer stand for (see Figure 9.2)? We believed that it was products that were user-friendly, dependable, and affordable. However, to reach this objective, had the company's people, culture, and attitude been adjusted accordingly? Acer's slogan is "We hear you," but to achieve this goal means that you must listen to the voice of customers and understand their needs. Only after you have established this foundation can you make customer needs the basis for the direction of your internal operations, while at the same time refining your message. In addition, since we are concerned with globalization, our message is a universal and consistent one.

RESOURCE ALLOCATION

Developing one's own brand requires at least the basic minimum of resources. I remember the "jumping off the cliff" idea—building a brand is like jumping off a cliff, and if you don't make it, you end up crashing at the foot of the cliff. The larger the scale, the bigger the goal, and the

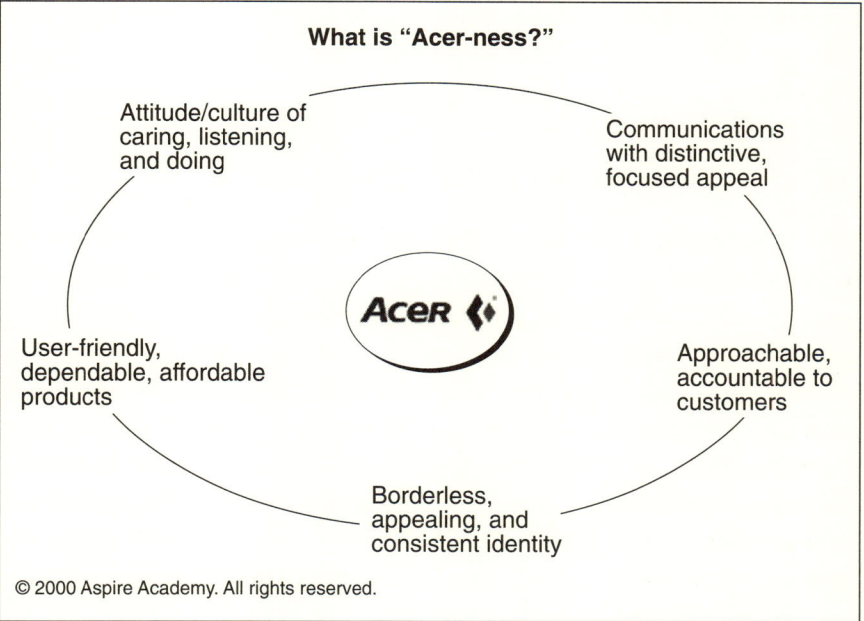

Figure 9.2 Acer's total brand management (2)

steeper the cliff. You must have at least a critical mass of resources, and determine the market segments, geographical regions, and products you are targeting; if you cannot ensure that you have at least the basic resources, everything will probably be all for naught. The most important thing is how to choose the cliff that you can "jump off" successfully: you muster the confidence and find an approach to make the jump, and slowly expand. When you're building a brand, you can set a large number of milestones. North-east Asia does not do business this way; it just thinks about getting the next order without doing any detailed planning. Most U.S. companies are different; when they want to build a brand, they determine a focus and then concentrate their efforts. In creating a brand, everyone always feels there are not enough resources, so you have to focus your energies, and during a particular time concentrate on a specific market or product.

I often reflect on why the same Acer product performs so differently in different countries—in Italy it's number one, but it loses money in the U.S. The fundamental reason is that the local management structure may have developed problems, and that the key is not in the product itself, but in management. So you must work in a focused manner, step by step, establishing the right conditions for effective management. From the perspective of management, when the size of a manufacturing team goes from 100 to 1000 people, it's very simple to make the adjustment,but it's extremely difficult to manage a sales team that has grown from 100 to 1000 people—it may require a very long time to make the necessary adjustments.

Channel considerations

In the international market, you have to go through the channel in order to build a brand. Some people believe that in the Internet age, you don't need the channel and can go directly to consumers, but I don't think things are that simple. In the online world, there are also online channels. The question is: do you serve as your own product's channel, or do you go through a virtual channel and work with another company? What's more, many products must go through real-world channels to really satisfy consumers. When products are to be sold on the international market, you have to think about distributors, retail outlets, and end-users in deciding exactly what kind of channel to build. Do you contract with an exclusive agent or use a multiple-vendor channel? If a product is very compelling, a multiple-vendor channel can be used; if the product is weaker, then you settle for an exclusive agent. In the early stages of Acer's entry to the international markets, if we did not grant exclusive agency rights, no one was willing to help us out. But when Intel came to Taiwan, it not only did not give us exclusive agency rights, but also wanted us to give up our AMD products. The business world is very coldly pragmatic, and everything is based on who has the power. Acer was originally AMD's biggest agent, but in order to do business with Intel, we had to give up AMD.

As Acer's brand gradually strengthened, there was a slow shift from exclusive agents to multiple-vendor channels. Besides this, when we went to Europe, we had no choice but to serve as our own distributor. Being one's own distributor also has many problems: you have to handle marketing expenses and inventory, support yourself, and work directly with individual businesspeople. When you have a designated distribution agent, or you work with retail operations directly, the biggest problems are after-sales service for consumers and inventory management—the question is: who should be responsible for these functions? In the U.S., Acer sold through large retail chains. This type of channel is a little like "try, then buy"; we send stock to them and they can return it, and we're responsible for anything that goes unsold.

Another issue is the nature of the relationship between the channel and the company: whether it is a loyal relationship. The biggest difficulty in selling PCs is the channel pipeline problem. Who exactly does the channel belong to? This involves two very different approaches: push and pull. In Taiwan, push is traditionally used, with products being dumped on distributors, who then dump them on retailers—since profits are very low, it's impossible to shoulder more responsibility. The U.S. uses the pull approach. The retail price is set by the company, the right to lower prices also resides with the company, and the company has to take responsibility for any unsold product.

The biggest headache with PCs is the "me-too" mentality; products are all alike, so everybody has to fight over channels. The result is that PC vendors have to shoulder all the burden. Up to now, all the PC companies have lost, but they still can't give up on this channel because brand image is like a kind of momentum, and to fight a battle you need momentum. If you don't have products at retail outlets, a huge gap in your market presence is created. After Acer exited the U.S. retail channel, we lost a lot of our previous momentum. However, it would have been extremely expensive to stay in that market in order to maintain momentum, so for the time being it is better to cut losses even if it means a loss of face. Later, IBM also exited; HP could not: it still had its inkjet printers and scanners in the retail channel, so it was stuck.

Lines of credit for channel vendors are another source of trouble. In the early years, there were several Taiwanese computer companies that collapsed because channel vendors in Europe or the U.S. didn't honor their debts, and this is especially likely to occur when markets are in transition. In 1991, when Compaq cut their prices by 30% or more, European channel vendors virtually all collapsed, either changing their business or going out of business entirely.

The approach we're taking now in mainland China is already in its third version—we've shifted from using a Hong Kong exclusive agent, then more than ten agents (some shipments made to Hong Kong, and others directly to mainland China), finally to our current system of going through 300 distributors within China, whom we are working with directly. No matter how stringent our approach is—requiring letters of guarantee or payment in cash—in the end, when the market is growing, distributor resources are always insufficient, and when they don't have money, they want to negotiate with you and want you to let them delay payments. However, as soon as you start doing that, the amount accumulates, and finally there's no way to recover it. At the time, I did not agree to payment delay, but people directly involved had a hard time refusing them because the market was finally opened and there were so many orders. So they agreed to give out some lines of credit and then started having to deal with non-payments. Compaq lost more than $100 million in China, which they won't be able to recover. If you give distributors a chance to take advantage, they will want you to increase their line of credit: it's so much faster to not pay than to actually earn money.

As for doing joint sales promotions with a channel vendor, unless the vendor is an exclusive agent, it won't be willing to work with you. If the vendor is not an exclusive agent, you can use them to promote sales, but you'll have to spend your own money to do it—just as when Intel uses all PCs as a channel, the expenses of the "Intel Inside" promotion are mostly borne by Intel.

CUSTOMER SERVICE CONSIDERATIONS

In the area of customer service, do customers actually belong to the manufacturer or to the company with the brand? Only with a brand can there be customers, and the identity of the actual manufacturer is not significant. The key to building a brand is service, and the simplest approach is to have a product that doesn't require service. Tennis rackets may not require much in the way of service; bicycles require somewhat more; and computers require a lot of service. Finding a product that doesn't require service may be the decisive factor in whether you can establish an international brand. Monitors require much less service than PCs. The effort needed to provide services cannot be taken lightly; even providing service in Taiwan is a huge challenge.

Providing services in the U.S. market is extremely arduous. Consumers will sue because they believe their hands were injured by using a keyboard; IBM, Compaq, HP, and Acer all are regularly sued.

In the U.S. and Europe, there is also the problem of third-party service providers. In 1990, when Acer entered the U.S. market, we found that providing service was an enormous drain on resources, so we hired another company to provide services for us, but this didn't work well either because we still had to provide back-end support, personnel training, and so forth, and customers still made complaints directly to us. Even though we used a service network that was already deployed very formidably in the market, the responsibility for service quality belongs to the company with the brand; consumers don't care whom you've contracted to be in charge of service.

Conclusions

Whether an Asian company should pursue OEM or OBM is a difficult choice. Without preparation for a protracted struggle, OEM is a viable choice. OEM is by nature an unstable business, but besides manufacturing, Taiwanese OEM companies can also do design. They have invested much in this area, and they must shoulder all the risk, but since they can do more, their level of competitiveness is greater than before.

After carrying out OEM for a few more years, a company gradually accumulates the capabilities to go the OBM route. In the information technology industry, Taiwanese companies' ability to build their own brands is much stronger than ten years ago, with the main reason being the bargaining chips they've accumulated in the process of doing OEM and ODM. However, there's no hope of doing this with old products because there is no innovation. The risk in making new products is very high, so this is also a dilemma.

OBM requires a different development approach, with the ideal being the use of a niche product to open a whole new market. An example is Trend Micro, which started in Taiwan and has done well in Japan and the U.S. However, does Trend Micro count as a Taiwanese company, or a Japanese or U.S. one? In 1978, I had the notion of moving company headquarters to the U.S., because in order to build a brand, having headquarters in the U.S. is a very big advantage. In fact, years ago there was a company called American Research Corporation (ARC) that was the U.S. headquarters of a Taiwanese company, and in the early stages it did in fact gain some benefits from this approach.

The most important thing for Asian companies is that they must have a large share of the mainland China and South-east Asia markets. If they can't, who can? The U.S. doesn't have the advantage of proximity that we have, and in battle, the army fighting far from home is naturally at a disadvantage. In building a brand, companies can take advantage of the U.S.—whether by getting coverage in industry magazines or winning awards in competitions, the U.S. can be used for building a strong public image.

It should be emphasized that technological development is global, while marketing and services must take into account local factors. In terms of long-term business models and future developments, OEM and OBM can coexist, and companies have to give both their due. They both present business opportunities, and they share a large common base of know-how. Even a brand such as IBM has started doing OEM, and Japanese brand companies have also started doing OEM. From the perspective of operations, looking at my smiling curve graph (see Figure 1.4, page 11), the left and right sides have already split from each other. In the past, joining the two sides together was standard practice: technology on the one side, and marketing and services on the other were integrated. Now though, all lines of businesses are separated. Technology should be like a multiple-vendor channel: the more the better; your own brand can take advantage of some technology you've developed, and OEMs can be given special licenses to do the same. As for marketing and services, which depend on brands, careful attention must be given to them—you have to be in complete control. It is commonly thought that there is a conflict between doing OEM and having one's own brand, but as time goes by the distinction between the two will get more and more blurred. So this is very favorable for future development in Asian countries that have been strong in OEM business but less successful in establishing global brands.

Acer's building of its own brand is a significant act. Building a brand means moving decisively, but in hindsight it is apparent that in some areas, Acer moved too quickly. For example, we probably should have waited longer before entering the U.S. retail PC market, building up resources, creating more competitive products, and refining our management model. We could have enhanced our brand through marketing activities, without being too ambitious in terms of the range of products we tried to sell in the U.S. market because, particularly for non-U.S. firms, the more a company sells there the greater the losses.

Discussion

Q: **You mentioned that in going after markets in the U.S. and mainland China, you met up with some frustrations. Could it be that before going in, Acer didn't analyze and assess products and channels?**

A: In terms of concrete action, no. Taking the example of being sued over the size of 14-inch monitors, it was a U.S. vendor that first made the mistake and we just went along. Pursuing a market almost inevitably means doing as the local vendors do, unless your product has an absolute advantage and people have no choice but to come to you. For example, in doing business in Taiwan, we use letters of credit, and it is importers and purchasing representatives who come to Taiwan. After selecting the goods they use letters of credit to make payment. After shipment they have to take responsibility for any problems that occur: before the transaction they examined the materials, risks, and other factors. However, when we go to the U.S. and make shipments, we're not selling to an importer. We have many buyers, but if they buy goods and there's a problem, they can return the goods. We request that they make up a letter of credit, and they ask "What is a letter of credit?" or say that the people in finance are not able to make one up as excuses for not providing one.

To build your own brand in the hope of penetrating a market, you have to follow local business practices. However, in different areas of the world, there are so many different customs that it's impossible to understand them all, which leads to frustrations. I always stress that Taiwan must train people in this area and must study markets to understand their rules.

Just as when we go into overseas markets, we get taken advantage of, so when foreign companies come to Taiwan, they get exploited; this is just part of doing business. For example, Taiwan uses a type of long-term financial instrument called a "bamboo rod check," which can be used for cash transactions in Taiwan. When foreign vendors come to Taiwan and are told they must pay with "bamboo rod checks," they're done for. It's the same for us overseas; if you're not careful, you get beaten up.

Use the China market as a training ground for going global

Q: You constantly stress that you want to be strong in mainland China, but China is ruled by personal relationships more than law; "paying your dues" may not result in gaining good experience. Is there an effective approach for doing business in mainland China?

A: Acer's investment in mainland China accounts for less than 3% of the group's total investment. We need to assume that because of its market economy, China will gradually adopt the rule of law. The China market may become the world's second largest, and many traditional markets, such as televisions and food products, may even become bigger than their counterparts in the U.S. Setting aside political issues, should you not go after such a large market? If there's an opportunity that you can perform better than U.S. companies, is there any reason to give it up? If Taiwanese companies want to build their own brands and don't take advantage of the mainland China market, they'll suffer for it in the long run. European, U.S., and Japanese companies all place a great deal of importance on the Chinese market. They have already been successful elsewhere, and all that's left is this market, which they couldn't get a piece of before because it was closed; now it has become the prize that everyone is fighting over.

The road that Taiwanese companies are taking is the opposite. We need to start from here and get into the world market. We must first use mainland China's market as a training ground and establish the requisite economic scale and cultivate enough professional talent; only then will we be able to compete in Western markets on an equal footing with foreign vendors. For Taiwanese companies, mainland China's market represents a training ground for worldwide markets.

In the PC industry, unless the country has no indigenous PC manufacturers or is too underdeveloped, the leading vendor in its market is never a foreign company. In Japan, Korea, Taiwan, and the U.S., the number one vendor is a domestic brand; there is no single international brand with an absolute lead.

We cannot put everything into the mainland China market and ignore other markets—we do not fight battles we cannot afford to lose. From Acer's point of view, taking investment in mainland China to 5, 10, or even 15% of our total would not be excessive; even at 20% it could not be called too big a risk. We must aggressively go after the mainland China market; the only caveat is that until we have established the needed management infrastructure, we won't rush in blindly.

Most Taiwanese companies that use mainland China as a base for pursuing global markets will most likely enhance their competitiveness. Outside of notebook computers, Taiwan's current information technology industry does more than 90% of its manufacturing overseas, and the majority of this is in mainland China. Taiwan's information technology industry has already completely integrated into mainland China and reached a world-class level of competitiveness. Larger Taiwanese companies probably all go through appropriate channels to establish a mutually beneficial relationship with the local government. As a result, it encouraged some small and medium-size businesses to set up satellite factories; this is a positive development.

The biggest problem today with doing business in mainland China is the legal situation found there. We are excluded from much business because the government will deliberately use ambiguous laws that favor local companies as part of a protectionist scheme. In fact, there's no need to be too pessimistic: mainland China has progressed greatly in recent years, and the business environment is getting better and better. You have to think it over very carefully, and go forward step by step. If the political situation shows any negative changes, you can get out any time.

Managing a brand is a permanent commitment

Q: Acer has a definite positioning for its brand. How can you convey this positioning clearly to end-users so that you can achieve the desired effect?

A: Managing a brand is like quality control: you have to put a lot of effort into it and constantly pursue improvement. We have not set any specific target, but each year we will evaluate the progress that has been made. If we run many campaigns, but they are not very effective, we may need to change directions. Should we use public relations, or public service events? There are different approaches. We will probably set aside a certain percentage of the budget for marketing expenditures, and then plan some approaches that will be more effective.

Do you have a fixed approach to total brand management? Is it consistent? Is there a consensus? If the answer is no to all these questions, does it make any difference? We evaluate whether something is worth doing by considering these questions. The U.S. many years ago realized the importance of strong branding, but only recently has this sort of total brand management system made its appearance. In the area of quality control, there is much talk about the so-called six-sigmas concept—domestically, it's seldom used—that advocates looking at your business not only from the point of view of product quality, but also from customer value. These things are for fine-tuning the management side; without them you can still get by.

Q: **Acer is a success story in brand building. Should other companies in the information technology industry try to create their own brands or start by doing OEM?**

A: In relatively mature industries, we have less of an opportunity to build our own brands. When Acer started building its brand, the industry was still not mature; we started with the Micro Professor series, and it wasn't until our fourth or fifth product that we finally started doing PCs. If the PC industry had already been mature, we wouldn't have had time to slowly make our way. Before an industry segment is mature, you can use innovation and a long-term vision to gradually create your own brand. After establishing your own brand, and you are still willing to do OEM, people will still work with you. If you first do OEM and then try to build your own brand, it may be considerably more difficult and even impossible.

Semiconductor companies also need to build brands

Q: Is the main brand or are the sub-brands more important? The customers for most components suppliers are not consumers; do these companies also need to build their own brands?

A: Companies in the semiconductor industry also need to build their own brands. The Intel Inside campaign was also aimed at building a brand. Every company has a name, and it should be integrated with its actions and image. If a brand is able to make consumers think of you when they contemplate whatever it is that you do, that lowers costs. OEM companies can also make it so that other companies seek them out when they think about what it is the OEM company does—that way, doing business is much easier.

From the standpoint of brand, in general you should focus on the main brand; but for fad-type products, because each product has a different positioning, trying to get blanket coverage with a single brand won't do. Trendy merchandise requires the use of multiple brands, while cars, computers, and machinery all need a single brand. In the long term, Acer has only two sub-brands, one of which is Aspire and the other is Acer121 for Web services. I use the Aspire sub-brand to take advantage of its reputation for innovation to enhance Acer's image in the U.S. because Acer is just an ordinary PC brand. There, I use "Aspire by Acer" to take Acer to a higher level. In Asia, we use "Acer Aspire" because "Acer" already carries a strong brand image here.

Part 4

e-Leadership

CHAPTER 10

Soft Infrastructure—Vision and corporate culture

CORPORATE CULTURE is a company's most important "soft" infrastructure, and it is also the key to a company's success or failure. As a business pursues excellence, the most basic qualities it must possess are a great vision and culture.

WHAT IS MEANT BY "VISION"?

What is vision and why is it important? How can a company develop an effective vision that everyone is willing to work toward? With the changes and growth in scope that time brings, a company must often modify its original vision and put forth a new one.

A vision is a kind of realizable dream. Taking the idea that Taiwan must strive to become a "technology island" as an example, the government's investment in the Hsinchu Science-Based Industrial Park can be considered a bargain—its investment was very limited, and yet it was the catalyst for the investment of hundreds or thousands of times more human and financial capital by Taiwan's industry. From this precedent, we can see that a vision can stir people into action.

A vision is not simply a goal, but something that is worth everyone's long-term effort. Becoming world number one is a goal, but not a vision; if you become number one, what then? A vision is something you pursue constantly over a long period of time, and not something you can achieve by tomorrow; however, it can't be something so distant it seems unattainable. Few will have the patience to pursue something whose end they cannot see, a vision they have no confidence in. The direction expressed in a vision must be correct. In establishing a business, there must be some idea behind it, a mission; and a vision should be related to the mission if it is to be effective.

Why vision is important

The reason that a vision is important is that it creates a sense of mission, making everyone willing to dedicate their complete effort to it. A vision provides cumulative results; a business is built up over time, and to become effective it has to encourage everyone to work in the same direction. To make the results of these efforts grow, a vision is needed, and no task should stray from the imperatives of the vision. This is especially true of running a company in the current business environment, which, although free, is extremely competitive. A company must be one step ahead of its competitors, and when they haven't yet sensed a change, you should have already developed a vision.

With the help of a vision to direct your efforts, when a market opportunity appears you will have already prepared all the needed competitive weapons—collectively called "entry barriers." Taking the Internet as an example, if a company lacks a vision and just imitates what U.S. companies are doing, the result is predictable. In the early stages, because it has no experience, the operations are not effective. Even if the company is able to achieve some early success, it has not built up an entry barrier to protect the business territory it has staked out; that is, other companies will be able to do the same thing at a lower cost and higher speed, and do it better. Therefore, when a company gets into a line of business, if it has no vision, there will be very great risks.

How to develop a vision

To formulate an effective vision, you must first get an accurate sense of likely future trends in the larger environment. The realization of a vision generally requires many years, which necessitates the setting of shorter-term goals to serve as signposts in moving along the direction set by the vision. In addition, you must have an adequate understanding of what you are undertaking; you must be clear about what is covered within the scope of your business.

In the main, a vision concerns the company itself and its business partners, and not its relationship to the outside world. A vision is the work of many people, and in general key executives should be sought out for discussion, with everyone contributing their thoughts through brainstorming sessions and creating a vision for the entire company or organization. After key executives have come to a consensus about the vision, you must communicate from the top down to everyone and form a group consensus.

A vision is at times just a direction, and it can be interpreted in many different ways and may change with time; because of communication, it may become progressively clearer, and gradually be transformed into practice. This will actually affect the way people work. The various ideas that exist within an organization are really linked to each other. For example, everyone in the Acer Group shares in the vision of fresh technology, and this leads to the vision behind the Acer brand of "breaking down the barriers between people and technology." These two visions build on each other; for everyone to be able to enjoy fresh technology, the barriers between people and technology must first be broken down.

Three years ago, Acer as a hardware company entered the software business, and in doing so we came up with the vision of "human-touch bits." In fact, this vision and the Acer Group vision of "breaking down the barriers between people and technology" are really just two sides of the same coin: everything that Acer does in the software business should be something that everyone can take advantage of. Recently, Acer has gotten into the Web services business, with a vision expressed as "the Web lifestyle—making us shine," meaning that we aim to enable the Web to make everyday life better. The organization as a whole, as well as subgroups within the Acer Group, is moving in the same direction, and this makes it easy to focus strengths.

WHEN A NEW VISION IS NEEDED

When does a vision need to be reformulated? When there is a change in the external business environment, a new vision needs to be sought. Revolutions caused by technology, such as when the engine or the Internet was invented, inevitably lead to drastic changes in the business environment; therefore, companies must come up with a new vision in response. Acer's "fresh technology for everyone to enjoy" idea is meant to make the company's vision more "neutral"—even if new technology appears, we can still enable people to enjoy its benefits. When the external environment changes dramatically, it will not have too large an effect on more neutral visions.

If the changes in the external environment affect the company's direction, adjustments must be made immediately; otherwise, much work will go to waste. When internal conditions at a company change, a vision also needs to be adjusted. Twenty-four years ago, Acer's assets amounted to NT$1 million, and it had between ten and twenty employees. Compared to the present situation, with NT$100 billion in assets and 35,000 employees, everything was completely different. Naturally, a different vision is now required. A vision is a dream that can come true, and if you are small, you cannot dream too big; after you make one dream come true, you can then come up with a new one.

When the top leaders in an organization are changed, or when the business is in decline, excuses should not be sought; rather, shortcomings should be acknowledged and steps taken to rectify them. The company must re-evaluate its original vision to see if it's really what's required.

ACER'S BUSINESS PHILOSOPHY

What a business philosophy represents is a belief about certain matters; for example, I believe in the idea of being a good corporate citizen. A company must take an idea that it deeply believes in and establish a complete value system that becomes the driving force behind the company's commitments. With a belief in this business philosophy, when working to establish competitive strengths, every employee will have a common direction, which will make it easier to realize the full extent of the organization's powers. Staying true to this belief is the only way to ensure constant improvement in a company's performance.

Acer's most basic business philosophy is that customers come first, employees second, and shareholders third. This ranking must gain everyone's agreement, with actions in full accord, and become the idea behind the company. Business theories put forward in the U.S. tend to center on creating shareholder value. Acer's philosophy is to persuade our shareholders that Acer does not operate for the short-term benefit of shareholders, but for their long-term benefit. To reach this objective, you must give employees peace of mind and serve customers well—profits will come naturally. Profits established on a base of benefit to consumers are the only sustainable ones, and if shareholders don't see the logic of this argument, I remind them that I am the group's biggest shareholder.

Acer believes that only through the complete sharing of know-how can an organization maintain its vitality. The traditional Chinese way of doing business is to withhold certain information as a way of maintaining power or a tactical advantage, so there's no way to constantly improve because wisdom is not freely shared, and the organization cannot develop vigorously.

Another keystone of Acer is distributed management, which does not only mean empowering others, but also relinquishing power. Everyone in society seeks to acquire power because the conventional wisdom is that it is easier to get things done, but I constantly stress the advantages of relinquishing power, explaining the notion that for many dreams to come true the dreamers cannot be subjugated by a single power. A strong dreamer has wisdom, so why does he or she need to be ruled over? An organization's best leader has a compelling vision, a philosophy that provides a direction for people to follow. Distributed management has its share of problems, but when the external environment is changing on many different fronts, distributed management that is implemented well will undoubtedly work best.

Another of Acer's basic philosophies is that it constantly weighs the state of its system of shared interests. If a team does not have shared interests, it won't work. So we implemented a company-wide employee stock ownership program, making the relationship with employees a long-term partnership.

How to create a company strategy

After some experimentation, we have established a comprehensive and effective procedure for formulating company strategy. The participants at a strategy meeting are leaders and first-rank executives, around ten people in total. More than twenty people is just too many, while much fewer than ten is not enough. In this meeting, everyone shares ideas and talks about big trends in the outside world, as well as the significance of these trends. Furthermore, we set out some directly related topics for discussion and gradually form a vision, or mission, for the company's future; the simpler the conclusion, the better—best of all is if it can be encapsulated in one or two sentences. Then, we must fix three-year, five-year, and ten-year objectives and think about what the key success factors are for meeting these objectives. We do a SWOT analysis, decide on some strategies, and, lastly, determine an action plan. During each step, we hope to come up with at least ten proposals, and after that everyone casts their votes.

ACER'S STRATEGY

The "peripheral attack" strategy described in earlier chapters is a very important part of Acer's approach. I constantly stress that many strategies are created by thinking about inherent weaknesses in the nature of the company—weaknesses that at least in the short term cannot be remedied, such as having limited resources. Because Acer's resources are limited, "peripheral attack" is a strategy worth considering.

Many companies are very successful at first in their own niche, but the problem is that the company is doomed if it cannot grow; and if it wants to continue its growth, it has only two choices: go after many niches or win over a major market. However, to pursue many different niches requires very diverse management capabilities, while to pursue a major market, you must have the ability to work in concert as a team. Whichever route is chosen, the required capabilities are not ones that a company typically possesses just after having started up and establishing itself in a niche. Many companies in Taiwan's Hsinchu Science-Based Industrial Park that went public in the late 1980s did very well early on, but ended up failing.

When continuously establishing new niches, using an approach of internal startups is more effective: it can resolve the difficulties of mixed management in a number of different businesses. The internet organization is based on the distributed management concept used by Acer, which gradually developed out of the client–server management structure. Though the idea of the internet organization originated with us, its objective was to re-engineer our organizational structure in order to achieve greater speed and flexibility in the face of extremely rapid changes in the business environment.

After settling on a strategy, the key is not simply talking about the strategy, but how to execute it. At that point, you need a philosophy, or a vision, to drive the strategy and effectively execute it. The creation of internal startups is a type of strategy, and everyone agreed with the need to train executive talent to lead these new companies. In 1990, Acer trained

100 executives, and in 1995 we presented a plan to train 200 more. When other companies were still putting forth their own spin-off plans, we were already way out in front: in performing these strategies, we had already accumulated a great deal of experience. The ultimate motivation for setting forth so many strategies was to erect entry barriers. As a company becomes successful, it naturally builds up certain burdens; if a large company has not built up some entry barriers to keep other companies from encroaching on its business, its success will be threatened.

DEVELOPING ACER'S SOFTWARE STRATEGY

In 1997, Acer sensed that the software business was becoming increasingly important not only for Acer, but would have an impact on Asia's future competitiveness. Software became a company direction, and we held a strategy meeting. Before the end of the first day, we had already come up with the vision of "human-touch bits." On the hardware side, we had only to come up with five-year targets, but for software we were stabbing in the dark and didn't have confidence. So we set our sights on a more distant future, an objective to reach by the year 2010.

At the time, the size of our hardware business was quite large, but our profits from software were virtually zero. We set a goal of one-third of profits and one-sixth of our revenues coming from our software business in the year 2010. Through SWOT analysis and analysis of the larger environment, we saw that it would be impossible to compete directly with U.S. software companies, so we formulated three directions. First, we would take advantage of our strength in hardware and develop software bundled with the hardware. The whole world's computer hardware companies depend on Taiwan, and if we produce software for bundling with hardware, we would gain an advantage. The second direction was to leverage our geographical location, do Chinese-language software content, beginning in Taiwan and extending to South-east Asia, avoiding Japan and Korea. The third direction was to provide local or regional services. The definition of the terms *local* and *regional* can be adjusted over time as conditions change. If one day mainland China becomes "local," then the scope covered by regional services can be expanded.

These were the directions set by Acer three years ago before it entered the software business. They have already become a basic philosophy, and whatever we do or whatever decisions we make, we don't go outside the boundaries set by these directions. We want to establish 100 companies, which implies that these software companies will not be large, with perhaps only 50 to 100 or 200 people, but will still be very competitive—this is a distinctive feature of the software business. We wanted CEOs who are less than forty years old, and this caused a lot of controversy because it immediately took many people out of the running; later we changed the principle by making "elderly in age, but young at heart" acceptable as well.

The corporate culture required in the software business is just what Acer has strength in—its characteristic ability to operate as a network organization, to use distributed management, though we realized that we would have to make some adjustments in our corporate culture if we were going to attract and retain the kind of talent who could develop Acer's software business effectively.

As for investment in the software business, we immediately set up an NT$100 million software fund (in the hardware business, investments are measured in the tens of billions of NT dollars), which gained a great deal of attention from the general public. In April 2000, Acer held the e-Life Show, defining ourselves as "enablers of the Web lifestyle," and the entire group adjusted to this direction.

When we were formulating our software strategy, although the Web had already been developed, it was not in wide use. Only in the last one or two years has it exploded in popularity, enabling the development of our software business to quicken. The vision we had about software has enabled us to take advantage of the current popularity of the Internet. At the same time, we are using the extensive resources of the group to their best potential by deploying them in a better manner. In 1997 and 1998, we established a software center in Shanghai, with the objective of training various types of software talents.

With the directions set at that software strategy meeting, each year our task becomes clearer, more concrete. By 2005, there may very well be 100 companies in our software business, and our hope is that they will be an important source of profits for Acer. In fact, Acer's main source of revenue now is the PC, but it accounts for less than 50% of profits; Acer is already being transformed.

How to cultivate corporate culture

Corporate culture is a type of value system; it is the invisible foundation of a company. For a company to effectively communicate its values, it should make use of slogans, but most importantly it should demonstrate its culture through its actions. A company's leader plays the most important role in the creation of a corporate culture. Many companies lack a corporate culture, and this absence is itself a type of corporate culture, signifying ineffectuality. If a company does not create a corporate culture, there is still the possibility of short-term success, but there is absolutely no chance of long-term success.

To create an effective corporate culture, it's best to start when the organization is still very small, using slogans, ubiquitous descriptions of the nature of the culture, and, if possible, some anecdotes that illustrate its application. Each person's behavior must be compatible with the interpretation given of the company culture, and there must be constant communication; that is, the creation of consensus. There should also be incentives, the means to reward those employees who exemplify corporate values.

The first type of philosophy I tried to inculcate at Acer was a belief in the inherent goodness of people, and this is a value I believe in (see Figure 10.1). In a more cutthroat environment, it might have been difficult to establish this sort of value, but when the organization was still small, it was easier.

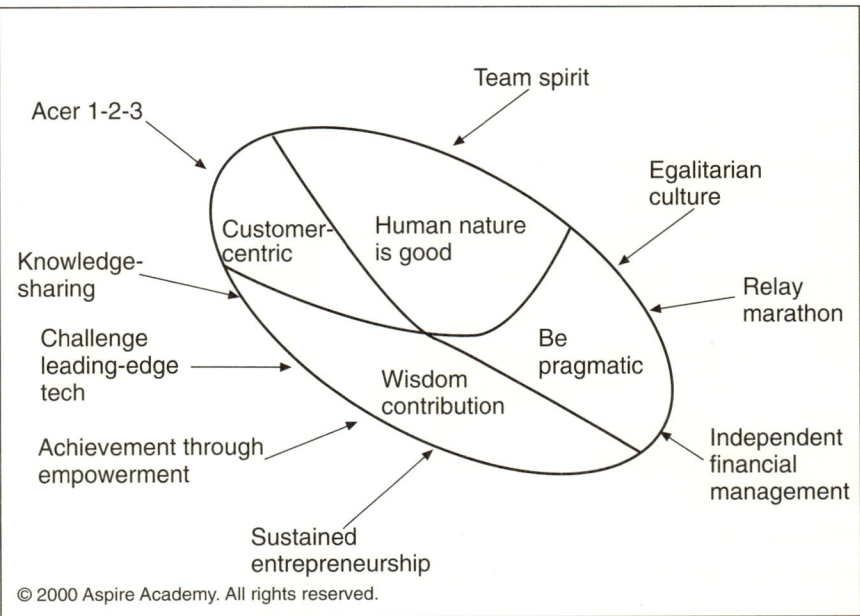

Figure 10.1 Acer corporate culture

A corporate culture that changes at the slightest provocation is not a culture at all, and a company without a strong culture will be overwhelmed by outside influences and eventually lose itself. Acer's U.S. operation lacked a corporate culture: the people there were too heavily influenced by external forces and simply adopted the commonplace values of U.S. businesses. Holding these values and trying to develop Acer's business not only made it impossible to demonstrate Acer's distinctive features and advantages, but exposed its weaknesses.

Acer has developed a business philosophy based on an egalitarian culture, emphasizing the model of a relay marathon and the achievements of people at every level of, or playing differing roles within, the organization. We hoped to be right at the cutting edge of technology and that finances could be independent and autonomous, with each company within Acer supplying a large portion of its own capital.

In 1980, Acer personnel got together and composed a company motto: "Benevolence, pragmatism, contributing wisdom, putting customers first." Previously, the group had always wanted to use my answer to the question "Should manufacturing lead, or should R&D lead?", which was, "Wisdom leads." During this process of cultivating Acer's corporate culture, I constantly emphasized these values.

Corporate culture is strongly related to competitiveness (see Figure 10.2). The notion that people are inherently good gave rise to conditions in which people at Acer could fully develop their potential. Respect for the customer means that we think from the customer's perspective, considering how to create value or customer benefits. This is also a part of my competitiveness formula. On the one hand, contributing wisdom is about creating value, and, on the other hand, it is also about lowering costs. Pragmatism is related to considering customer interests, while also controlling costs. Whether a corporate culture is sustainable for long is definitely related to the level of competitiveness it can achieve.

THE CHALLENGES OF ESTABLISHING A CORPORATE CULTURE

In 1990, I invited McKinsey Consultants to perform a diagnosis of Acer's operation. At the time, company sales had begun to decline. They pointed out that for a corporate culture to be effective, it had to be established when the company was developing smoothly. When sales are not good,

Figure 10.2 Acer culture and formula of competitiveness

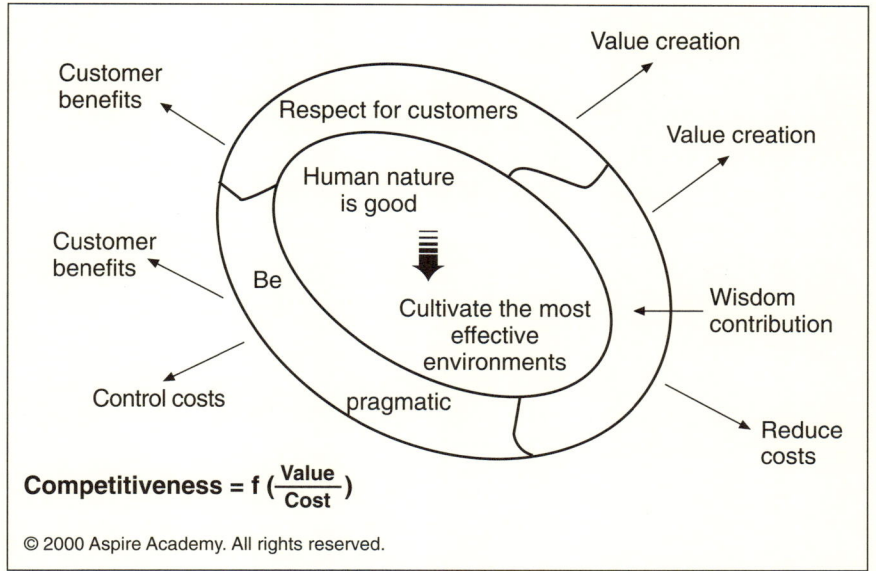

Competitiveness = $f(\frac{Value}{Cost})$

© 2000 Aspire Academy. All rights reserved.

no matter how good a corporate culture might be, it has no persuasive power. When a company shows strong earnings, the opportunity can be exploited to emphasize that the corporate culture in effect is correct; during times of success, you must prevent the appearance of short-term, misleading corporate cultures.

When an organization is growing rapidly, it is also difficult to cultivate a corporate culture, which takes shape slowly over time. The situation can be likened to a coal furnace: when the fire is strong, if you add too much coal all at once you'll put out the fire; slowly adding coal is the way to make the fire grow stronger. A corporate culture is similar: if an organization is growing too rapidly, it's very disadvantageous for the building of a corporate culture.

After a company has gradually diversified its business, it is also very difficult to create a corporate culture. In Taiwan, many traditional industries want to get into the electronics industry, but it's very difficult to create a corporate culture under these circumstances. A single corporate culture cannot be applied to every line of business; if a company from another industry attempts to enter the electronics industry, its original culture will lose its effectiveness. It's also difficult to establish a corporate culture in foreign subsidiaries, thanks to the distance between the company leadership and the actual operation.

CONCLUSIONS

At each turning point in a business's lifecycle, reflection must be given to the question of whether a new vision is called for, or whether the current vision needs adjustment. Company leaders play a key role in the formation of vision and business philosophy. Corporate culture is like marketing capabilities or manufacturing capabilities—it is a core competitive weapon. Moreover, it is a far-reaching and highly influential factor in the company's prospects; it is the basis for everything. With a vision, everyone can work toward the same direction; with a business philosophy, work can be sustained; an effective corporate culture can serve as a stimulus and source of motivation.

The corporate cultures of all outstanding companies share a common principle but manifest a different style, depending on the particular line of business, industry, or the individual leader. The book *In Search of Excellence* analyzed successful U.S. companies and classified their corporate cultures. It found that they were in fact largely similar. The interesting thing is that after eight years, most of the companies in the book were no longer performing as well: external conditions and company leadership had changed. A corporate culture must be completely assimilated by its leaders and executives at every level because business activity is constantly changing. If the actions of top executives do not accord with the company's vision, business philosophy, and culture, all the efforts made will go to waste. When running a business, one of the most important tasks for an executive is how to preserve the correct values within the company. Companies that do not pay attention to these issues but are very successful are very rare.

DISCUSSION

Selecting a culture that is effective for business

Q: How can a company's culture be transplanted to overseas operations while still assimilating the local culture?

A: When companies set up foreign operations, they should respect the principle of following local ways. However, the biggest problem that afflicts companies now is that they have not created their own culture even though they are already internationalized.

Ignoring overseas operations for the moment, a popular approach now used in running business in Taiwan is the "commission" culture.

If one is not conscious about it, this culture can naturally infiltrate an organization and become its default culture. A culture that is effective for one's business should be selected; and then through long-term communication a consensus should be formed, leading to values that most people can identify with.

There are several reasons why Acer America was not able to create its own culture—in effect, it had no corporate culture at all. First, when it was successful, it grew too rapidly and missed a great opportunity to nurture a corporate culture; second, employees were affected by external influences more than influences originating from the culture at Acer headquarters in Taiwan; third, when development faltered, employees wanted to introduce corporate culture from successful U.S. companies, but these were not things that our actual capabilities would allow. In addition, headquarters did not have enough authority to forcefully disseminate an effective culture and set of values.

It must be repeated, a corporate culture becomes strong because it is effective and successful and enables competitiveness locally. More than twenty years ago, I promulgated ideas such as an egalitarian culture, the relinquishing of power, and people as inherently good; these were effective because we won with them, otherwise they could not have become Acer's culture. As I've said, corporate culture must be built up over a long period of time.

Sharing wisdom, coaching without withholding

Q: **Does competition among employees for promotions affect Acer's culture of empowerment?**

A: A way must be found to integrate the culture of sharing know-how with a fair and effective promotion system. For example, when considering promotions, we will also take into account who will take over the promoted person's position; when selecting a leader for a key position, we will choose a manager with many talented people under him or her. Therefore, the promoted person must be able to relinquish the powers belonging to the original position and leave them to his or her successor. If you believe in this sort of culture, if by chance there is some flaw or blind spot, it must be compensated for through the system itself. When people are in a competitive environment, it's hard to completely avoid concerns about putting yourself at a disadvantage, and this tempts people to try to hold on to power, to withhold know-how. It is only through actual behaviors, such as at meetings or when people are working together—with people showing a willingness to fully share know-how—that this kind of corporate culture can be firmly established.

Convey the vision from the top down

Q: **Is a vision best if formed from the top of the organization down, or from the bottom reaching upward?**

A: If it is from the bottom up, with masses of people talking, how can so many views effectively coalesce into a single vision? When holding a strategy meeting and presenting a vision, the CEO leads the discussion with top-level executives. These people represent various functions and levels within the company, and they also understand the views of employees under them. They discuss issues related to larger trends, customers, and employees, and then jointly formulate a vision. Then from the top down things slowly unfold, and the vision is constantly propagated. When conveying a vision, one should have a mechanism for providing feedback; if there are misunderstandings, these should be clarified.

Acer's corporate culture is naturally led by myself; however, I gather many ideas on which there is consensus before coming up with the tenets of that culture. What I am thinking in my heart is always of other people's interest, even people outside Acer in society at large. When I came up with the internal startups initiative, I had already come to a deep understanding of what younger employees were thinking. They hoped to have their own businesses with an international stage to show what they could do. I don't need to ask every person in order to know this information; I already had a good grasp of the attitudes and needs of the people at Acer.

Making decisions is the responsibility of people higher up, but when they do come to a decision, they should already have incorporated the views of the people below them. Executives at each level represent different functions and voices; through the sharing of thoughts, a consensus can be reached. When promoting a vision or strategy as formulated, they can demonstrate their understanding of employees. In 1994, when I presented the "21 in 21" and "2000 and 2000" (with revenues of NT$200 billion in the year 2000) targets, this was not simply a matter of making up some numbers. What these targets represented was people having their own companies that could go public independently.

When a company is successful, create a culture

Q: What is Acer's strategy for nurturing and recruiting talent?

A: As mentioned previously, the creation of a corporate culture requires taking advantage of an interval when the company is particularly successful, and the same is true for attracting talent. In our previous experience, highly paid employees are not necessarily very successful: their starting point is not a long-term sustained effort. In the future, we will rely on an approach that does not depend only on spending large amounts of money, but one that is long-term and sustainable, to cultivate talent and build a corporate culture.

As for training talent, it is my hope that through the application of technology and the network, I can make the experiences and business philosophies I have accumulated available to employees. During the first internal startups initiative, I lectured to the point where I almost had to have throat surgery. I go to Acer offices all over the world to describe my business philosophy. Our employees in Asia have absorbed this; our employees in Europe and the U.S. have been persuaded. However, once I leave, the lesson is forgotten. What I talk about is mostly vision, business philosophy, and strategy—things that are more abstract. The ability to formulate and communicate these things is the key strength that corporate leaders and executives must have.

Senior employees recruited from outside must take a low profile

Q: How can executive talents who are suited to a company's corporate culture be chosen? What can be done to help them assimilate into the existing corporate culture?

A: This is not easy. Most U.S. companies do not have this problem: with a new CEO you throw out the old rules and follow the orders of the new leader. In Acer's experience, a senior employee who is recruited from outside must as much as possible take a low profile. There should be a considerable length of time when he or she should avoid trying to be a one-person show. The executive can be responsible for many tasks, but there will be someone working close by to help him or her assimilate into the corporate culture.

However, a corporate culture is an intangible thing; on the one hand, we wish to consistently uphold some principles, and yet on the other, we want to encourage differentiation. This is because leaders, industries, and businesses are all different and will give rise to different cultures. Principles seem self-evident to those who are accustomed to them; Acer has its way, and it takes some time for new employees to grasp the nature of the Acer way. So at Acer, it's very difficult to have an opportunity to demonstrate your abilities in a spectacular fashion right from the start. A certain amount of time is required before you can start doing things with a big impact.

Taking Acer Communications & Multimedia (ACM) as an example, ACM will assign people from inside the company to serve as general managers at subsidiaries and help a senior vice-manager recruited from outside the company understand the corporate culture. Acer has several subsidiaries where someone from the outside came in to run the show, and it caused a lot of trouble.

It's difficult to succeed using size to conquer size

Q: Right now, corporate mergers and acquisitions are popular; how can one's own corporate culture be transplanted into an acquired company?

A: To acquire another company, you have to be very strong; it's like a fire that has to be strong enough to sustain itself when a lot of fresh coal is added. There are very few instances, especially in the fast-moving information technology industry, of a successful acquisition by one large company of another large company. If the companies in a business group do not share the same corporate culture, there is no way to run it well.

Taking Cisco as an example, it has been able to acquire many companies—it is big enough to quickly assimilate them. It has two methods for assimilation: first, the information systems must be switched to Cisco's; the other method is through corporate culture. Cisco will post some of its own people to the acquired company to forcefully change its culture and integrate it into Cisco's. The most important things about corporate culture is that it must be big and powerful in order to "take over" another company. At Acer America, an outside culture ruled over Acer's culture, and this just did not work.

A dominant vision, corporate culture, and business philosophy are needed as the foundation for competitive strength. They become progressively more important: the more intense competition is, the larger the scale of a company, and the greater the rate of change in its industry. However, many companies when they are starting up don't face this kind of intense competition, so they can succeed without a distinct corporate culture. To achieve continued success, there are two choices: first, to create a vision, business philosophy, and strategic capabilities; and second, to find a niche in the market that does not require these capabilities for survival; this way you can survive, though you will not make any distinctive mark.

CHAPTER 11

Virtual Dream Teams

THERE ARE MANY TASKS that can only be accomplished by relying on an organization, with various kinds of tasks requiring organizations with different characteristics in order to achieve the best possible results. If an organization is to successfully accomplish the tasks with which it is charged, it must also cultivate talent and constantly develop new personnel resources. As an organization is situated within society, a company must adapt to the many changes that society is undergoing, which are making the range of tasks that companies must deal with ever more diverse. The virtual dream team approach is an effective means of providing companies with the capability to prosper under these challenging circumstances.

DIFFERENT FORMS OF ORGANIZATIONS

Whether an organization is rigid or flexible is related to its particular mission. If a mission is relatively fixed, and if the tasks it performs are repetitive, needing discipline, then using a rigid organizational structure is more appropriate. If the mission is constantly changing and requires innovation, along with the ability to take advantage of new opportunities, then a flexible organizational structure must be adopted.

Organizations with a traditional layered structure are relatively fixed because a large scope is involved; it's more difficult to make a large organization highly flexible. An army is a hierarchical organization; from the commander-in-chief down to rank-and-file soldiers, there are numerous layers. An army's mission amounts to one command corresponding to one action, and action must be consistent; the people lower down cannot have their own ideas. In this kind of organization, management is comparatively simple; normally, a great deal of training is required. A company's manufacturing operation is similar, with a great number of levels, since manufacturing requires discipline and is repetitive.

An organization with a flattened structure can be contrasted with a hierarchical organization. A flattened structure is a very important conceptual approach in management science: when company objectives are changing, and tasks must be passed down the management structure for handling, such a structure is more conducive to effective communication. Whenever a task passes one level, it loses something—its authenticity, its original emphasis—something that the fewer layers in a flattened organization minimizes.

In 1990, Acer made some momentous changes, with the first priority given to flattening its organizational structure, reducing the layers from seven to five. Over the previous ten years, Acer's annual revenues had increased by a factor of 1000; but because we lacked experience, as the organization grew larger, layers were added. I like to change, and colleagues often kid me, saying I'm the dragon's head, but the tail of the dragon can't keep up with the movement of the head because the head is moving again before the tail has a chance to respond to the first movement. While it's an amusing comment, it does point out that communication within Acer is not as efficient as it could be.

The form of an organization is related to the conditions in the external environment, and because of conditions in the past, most organizations have been centralized. We've progressed from an industrial society to a knowledge society, one of whose distinguishing characteristics is that tasks and missions constantly change—with the result that centralized authority is gradually giving way to more delegation of power, and hierarchical organizations are being transformed into network organizations. Should an organization be rule-governed or people-governed? If what is to be managed is a fixed and repetitive task, a rule-governed approach is appropriate; for example, in employee evaluation and promotions. The more advanced an organization is, the more tasks it must perform; in performing repetitive day-to-day tasks, the organization can rely on a rule-governed approach and should be able to achieve quick results. There are some rule-governed approaches that are similar to people-governed approaches, with the direction everyone agrees on becoming a type of principle. However, an organization should preserve room for some people-governed approaches and constantly utilize innovation to move the organization forward.

Totally rule-governed and totally people-governed organizations are both flawed. Totally people-governed organizations, without reliance on rules, cannot accomplish much. A network protocol is an element of a rule-governed system, but with this rule many different patterns can be created, many different things can be done, and a rich virtual world can be created—the same is true for network organizations ruled by protocols.

CHARACTERISTICS OF SUSTAINABLE ORGANIZATIONS

There are three real keys for creating a sustainable organization. The first is that the organization must retain an entrepreneurial spirit. During Acer's tenth anniversary, I said to all our people that we were only just starting our business, even though at the time Acer was already the largest information technology company in Taiwan. I constantly emphasized that Acer was only just beginning, but this manner of speaking was not compelling: its effectiveness was limited because the company was very big already. How could I say we were only starting up? We decided to split the company into pieces and go back to the start, constantly spinning off new companies internally and creating something from nothing.

The second key is to accept risk. Most large companies are not willing to do this, but over time companies that are unwilling to take risks cannot create new business. We encourage everyone to accept risk and learn from the process. The third key is that even though we stress constant change, there are many basic philosophies that have not changed from day one until now. While insisting on a few unchanging principles, you can constantly change.

For an organization to sustain development, it must constantly grow and consider how to become a growth-oriented organization. An organization must have a good system for renewing itself; if employee turnover is too slow, it can lead to problems. An organization grows in order to accommodate the personal growth needs of its people: if people do not grow, morale will be low, and the organization will become unstable. The reason that an organization must be flexible is that as external conditions incessantly change, there is no way for an inflexible organization to adapt.

A sustainable organization is one that promotes learning and has the ability to adapt and renew itself. Learning within an organization is of two types: first, individual, with every person using learning to further personal growth; and second, organizational, with many people learning together and thereby enhancing the organization's capabilities, thus allowing it to develop its approaches and workflow. An organization must also be able to easily re-engineer itself: re-engineering is a normal and necessary task. Additionally, an organization's entrepreneurial spirit is very important; only by maintaining such a spirit can an organization give rise to completely new businesses and sustain its vitality.

STRATEGIES FOR BUILDING A GROWTH-ORIENTED ORGANIZATION

To guide the formation of a growth-oriented organization, its leader must have the willingness and the motivation, and it's best if he or she also has a vision. The vision of an organization's leader should include two directions: first, how to facilitate the personal growth of individuals. One must have strategies for educating, retaining, and recruiting talent for the organization. When Acer first began, it had little money and few resources, so we could not give employees high salaries. Instead, I devised the approach of employee stock ownership as a means of giving them a dream—one that they would be willing to stay and work toward.

The second direction comprising a leader's vision for the organization is a question of adaptation. If a leader has a vision, he or she will consider a good approach to change: an organization must change over time to adapt to changing circumstances. The reason there are many organizations that find it hard to change is that this issue was not considered when the organization was founded.

It's best if personnel training within an organization utilizes empowerment, allowing people to learn by doing their jobs. This should not be a burden for employees; this is a more practical approach. Encouraging internal startups is a way to allow talented people to do things on their own. It is only by allowing this to happen that you can really change things within an organization and take the organization to a higher level.

Empowerment is extremely important, but it is even more important for an organization's leader to constantly think about how to invest for the future. The heads of large companies often complain about brain drain, about the loss of middle managers, and this situation is the result of not having targeted people at each level of the organization for investment. Investment does not only mean training courses, but more importantly empowerment—allowing people to upgrade their capabilities by doing things on their own.

STRATEGIES FOR BUILDING AN ORGANIZATION THAT CAN RE-ENGINEER ITSELF

A business that is constantly re-engineering is normal—when a company is successful, re-engineering is needed; and when it meets with a failure, re-engineering is even more urgent. Re-engineering requires conditions for open communications, and an internet organization has greater flexibility and can easily perform re-engineering; when communicating, consensus is more readily achieved. Compared to hierarchical organizations, when internet organizations undertake re-engineering the shock to the system is smaller.

In communicating about re-engineering, it must be explained to members of the organization why re-engineering is being undertaken, what the new direction is, and what actions will be taken. Even more importantly, the common interests of the majority of people must be considered, otherwise people will easily find excuses to impede re-engineering. Successful re-engineering depends on the efforts of the organization's leaders; leaders at every level should have some ideas about re-engineering.

ACER'S EVOLUTION AS AN ORGANIZATION

When the company was first established, the hope was to introduce microprocessor technology from the U.S. and promote it locally. At the time, our resources were extremely limited, so we only traded and consulted and helped vendors to design products; moreover, we only covered the domestic market. We had already begun to use a distributed management system, and our basic philosophy took shape at this time. The founders of our branches in Taichung, Kaohsiung, and the U.S. each held a 60% stake in their company, and headquarters had 40%. This is somewhat reminiscent of the disintegration idea. During this phase, our annual revenues were doubling each year, and even more importantly, while earning profits with limited resources, we built up more advanced microprocessor technology and high-tech marketing experience.

In 1981, the company officially set up in the Hsinchu Science-Based Industrial Park (HSIP) and began doing its own research and development, as well as manufacturing, with sales focusing on overseas markets. Prior to 1980, we were agents for Zilog, which was doing better than Intel in Taiwan; in the entire world, only Taiwan was able to buck the trend, demonstrating how effective we were as agents for Zilog.

However, no matter how well we did, our market was restricted to Taiwan, so we chose some products to export, to take advantage of larger markets. In this period, the business opportunities abroad far exceeded those in Taiwan; we needed a large number of people to handle them, so the organization adjusted accordingly.

In addition, before the company went public, besides Acer Sertek we also invested in Acer Incorporated in the HSIP and Acer Peripherals (now Acer Communications & Multimedia), which had been established at that point. Employee investment in the three companies took place over the same span of time; on the surface, there really were three companies, but in fact they were operated as a single company. The objectives and interests of people at Acer were the same, making it easier to move people around. Everyone at Acer Incorporated and Acer Peripherals came from Acer Sertek. If in the early stages Acer Sertek had not used investment in new companies as a way of shifting personnel resources, today's Acer Group would probably be quite different.

Since Acer capabilities were growing exponentially, each generation of startups could take advantage of even greater opportunities. At the time, Acer was growing at a high rate; in its first ten years, it doubled revenues every year.

After going public in 1987, Acer undertook to aggressively globalize its operations. When Acer did export sales, its products reached every part of the world; however, merchandise export is not the same as internationalization. Internationalization means having your own people assigned overseas; all the functions involved in running a business— such as marketing, sales, warehousing, manufacturing, and so on—must be international.

During this period, I contracted Ying-wu Liu from IBM to return to Taiwan from overseas and had him introduce current thinking on multinational business organizations. Because the organization was too large, it used an approach that mixed SBUs and RBUs. An SBU managed R&D operations and manufacturing, while an RBU handled marketing. Even though both SBUs and RBUs were part of the same company, they were separate profit centers. From the point of view of headquarters, they were parallel operations. As this approach developed further, I discovered that even if the two were involved in the same task, with the SBU handling the front end and the RBU handling the back end, it was very difficult to establish shared interests between them. Their work also could not be effectively integrated, and as a result, overall growth slowed.

The goals for the larger organization were quite broad, but they did not convey a sense of importance or immediacy to individual operations and were vague. Even more troublesome was the issue of interests; the interests of a large organization may not be directly related to the interests of individuals within that organization. At the time, when the final fiscal results were calculated, the difference in the level of incentives between profitable and non-profitable operations was not large, leading to the company's first-ever losses, reaching the point where for one or two years we had to sell off real estate holdings to sustain the company.

In 1992, the company undertook re-engineering and introduced the fast-food manufacturing and sales model, frequently launching fresh products on the market. This approach later gave rise to some problems: by this time Acer had established thirty operational centers worldwide, and each required professional talent to effectively manage inventory. The key to doing PCs is parts management; whether this is done well or not has a great impact on the company, and it requires the expertise of specialists. However, we were not able to post a capable person at every single operation, with the result that even though the original aim of the fast-food business model was to reduce inventory, the mishandling of this approach led to too much inventory instead, so we finally gave up this approach.

The organization produced the client–server organization structure, and we encouraged both the clients and the servers within the organization to be independent entities with the ultimate objective of going public. We therefore came up concurrently with the plan to have twenty-one companies go public in the twenty-first century. Our original modest growth suddenly increased to an annual 50% or even 80% before gradually slowing again. The SBU/RBU structure developed problems: each RBU was autonomous and headquarters had no global strategy for products and marketing. As a result, Acer began its re-re-engineering effort.

In 1999, because the organization was growing larger and larger, subgroups began to form within the Acer Group. The current Acer Group is only virtual, with the Acer Incorporated, Acer Communications & Multimedia, Acer Sertek, Acer Venture Capital, and Acer Digital Services Group subgroups serving as command centers. All the SBUs and RBUs have been assimilated into GBUs, with the upstream, midstream, and downstream operations all complete in themselves, operating and interacting like nodes in a network. A GBU is not global in the geographical sense, but in the sense that each is an independent entity; and using this approach, 100 or even 1000 independent GBUs can be created. This kind of organization is not suited to a client–server structure, so the internet organization structure developed naturally.

The company's organization, originally a traditional business group made up of directly related companies, has now added franchises and outside startups funded with Acer venture capital.

Cultivating leaders for sustainable growth

For an organization to grow, it must constantly develop corporate leaders. A corporate leader must have an important characteristic: being able to accept others who are stronger than him- or herself. A corporate leader should be able to tolerate people with viewpoints or personal styles different from his or her own. Training a leader also means taking on new tasks from which to learn. A leader must constantly accept risks: failing to do so will result in an inability to make breakthroughs.

There is another important condition in cultivating talent: the company must have a corporate culture that emphasizes a leadership style of coaching without withholding know-how. Coaching without withholding know-how has two modes: first, the person higher up should demonstrate the principle, helping to establish the culture. Second, if an environment is not one in which this style of management prevails, individuals can take the initiative and help propagate the idea.

Characteristics of a successful corporate leader

A successful corporate leader possesses several distinctive characteristics, with the most important being the willingness to train people. For a corporate leader, training people may be even more important than doing one's own job well. To train other leaders means to constantly empower and help such personnel to achieve their tasks. A leader must constantly communicate—vision, business philosophy, and strategy are all important parts of a leader's work, and these things can only be understood through long-term interaction.

Especially in the case of tasks directly related to the mission pursued by a company, the people being trained must have a sense of accomplishment and involvement. Finally, the most pragmatic things of all are power and money; give the future leaders the power to make decisions, and give the monetary rewards as incentives to help them.

A corporate leader must also concentrate on key tasks; tasks that are less central should be delegated to other people or be handled by a rule-governed system, to allow the leader to focus his or her energies on key and new issues.

THE MOST COMMON PROBLEMS IN A BUSINESS

The problems that most business leaders develop are a lack of trust in anyone but close associates, a lack of long-range plans for cultivating new corporate leaders, and even an unwillingness to invest in personal growth. Buying a luxury sedan, having an impressive office, or earning more money—this is not personal growth. Personal growth is constantly having new challenges, new tasks; it means accepting risks.

Many businesses, after establishing a rudimentary foundation, are only willing to change a pattern of performing familiar, repetitive operations, and are not willing to enter a new environment and accept new challenges. Keeping to repetitive tasks may be advantageous in the early stages, but later supply will outstrip demand in the market because there are no barriers to entry, and this is a negative. After the same routine is performed for a long time, the return on investment will eventually decline. If a company does not place importance on cultivating talent, it will discover that people who have been trained for a long time all end up leaving to help its competitors, while the people who remain are mediocre.

HOW TO ENSURE A SMOOTH LEADERSHIP SUCCESSION

Training successors is an extremely important task for a company, but a successor cannot be trained in one or two days. It is a process that requires long-range, systematic thought and planning. The approach to successorship in U.S. companies is completely different from the one I advocate for most Asian companies. Talent is plentiful in the U.S.; the president and CEO can be changed and a person from outside brought in.

My approach is more useful in Asia because culture and values are different from those of the U.S. In order to reduce risks, companies must cultivate a group of candidates who are prepared to take over as successors. During Acer's course of development, there were at least ten people whose qualifications and skills were the equal of the current leaders of Acer's subgroups; it wouldn't have made a big difference whichever of them had been chosen. In the U.S., there are many people who are capable of heading any given company, so a U.S. company often looks outside itself for a new leader, while my approach entails conscious training of a new leader within the company itself.

The most important thing is that the daily work of a company is making arrangements for successorship; this is really what each day's work is about. Leaders have to constantly convey their own business philosophy, their view of things, to people who may one day succeed them. This should not only take place when the company is doing well, but it's even more important when the company is having difficulties; only in this way can everyone work better together and achieve consensus. To keep talented people, you must give them new tasks, autonomy, and a sense of belonging and achievement. It is very difficult to retain talent with just money.

THE ROLE OF CORPORATE LEADERS IN THE TWENTY-FIRST CENTURY

The twenty-first century will be dominated by technology, so if a corporate leader does not understand technology this will naturally lead to problems. What is meant by understanding is not the focused and deep comprehension of an expert, but having a grasp of the distinctive features of technology and the larger direction of future trends. Corporate leaders definitely cannot be afraid of technology; CEOs of high-tech companies rarely come from a non-technical background. The reason is that high-tech requires investment in the future and involves less repetition; one must seek constantly to understand new developmental trends. However, a person's college major is not the key consideration, but the effort made and experience gained after graduation.

The most important thing is that because the future will be dominated by intangibles such as knowledge, corporate leaders must understand the relative value of various intangibles. Value judgments of this sort are extremely difficult, but a corporate leader must have this ability: to see the value and influence of intangibles.

Conclusions

Cultivating talents requires three conditions: time, money, and the experience of those involved. A company must also supply a favorable environment, and this requires time and money and is not something that can be created at will. An internet organization is more natural and sustainable, but it is not easy to manage and it requires the relinquishing of power, making it difficult for the majority of corporate leaders to accept. Different organizations take different forms; this is related to the nature of their businesses and the style of their leaders.

An organization requires constant re-engineering, and a system for creating startups within the organization is an important factor in sustaining a company's vitality and making uninterrupted growth possible. There are many different approaches to creating internal startups: establishing a new company or just a new profit center are both examples of internal startups. In the future, the real and virtual will meld into each other; a real and a virtual company will both be companies, and perhaps both will be counted as startups within a parent company.

Discussion

Q: Corporate leaders should have integrity, talent, and tolerance. How are these three characteristics ranked?

A: There is no absolute answer. Looking at things from a long-term perspective, all three of these qualities must be present. As times change, in different industries with different teams, or at certain stages, talent might be the most important, while at other points ethical treatment of others and the ability to tolerate differences may be more important. A corporate leader must possess at least a minimum level of these three qualities; otherwise, while success in the short term may be possible, there will definitely be problems over the long haul because doing battle requires having all three types of weapons; lack one and it's your fatal weakness.

Use people who are stronger than you are

Q: Corporate leaders must utilize people stronger than themselves; however, in the actual operation of a business it's often discovered that these people have exaggerated their abilities, do not get along with others well, or are not willing to compromise. How do you deal with this situation?

A: "Strong" has different definitions: strong in skills, strong in negotiations, or strong physically. In utilizing people who are stronger than you in various areas, there is one area in which you will surely be stronger—that is, that he or she is being utilized by you and not the other way around; you are better at utilizing people. I constantly encourage the people under me and tell them that they are stronger than I because when I was their age, I had done less than what they have already done. If you let people stronger than you waste their potential or become complacent, it's a problem with your management.

I constantly stress that there must be many successors, and a single person must not be allowed to think he or she is stronger than everyone else. If a leader is devoted whole-heartedly to cultivating a talent and is not afraid of that person being stronger, that person will surely do his or her utmost for the leader if given an opportunity to prove him- or herself. If one day this person ends up above the leader, he or she will be more likely to treat the leader civilly. A strong person cannot be kept down, and if a leader tries to suppress a capable person, and this person does get the chance to succeed, he or she will become a threat.

The Acer Group goes virtual

Q: Is the Acer Group's preparations for leadership succession already complete?

A: The basic procedure for succession in the Acer Group has been set. I have turned the Acer Group into a virtual organization; there are not too many things to do. The work I do can easily be covered. All operations are taking place in subgroups. In the future, the group will work through committees, with everyone taking a turn as chairperson of this committee—a group of people taking the place of me, and not a single person taking over. There is no reason for one person to take my place because subgroups are the operational centers.

Every task must have someone to serve as leader: if there are two people in charge, and subordinates have to try and please two people, they will try to find a way to drive a wedge between the two. However, there can be many different leaders within an organization, each serving as the head of different parts, whether as small as a working team within a department, or a company within a business group. The Acer Group's biggest difference from other businesses is its extreme diversity; because it is like this, it needs many leaders.

The head of the subgroups are the corporate leaders, and the head of the Acer Group as a whole is like a totem; this is already what I am. When I think about these problems, I look at management from the point of view of natural principles, an environment in which people can thrive and that can make the company able to sustain itself forever. When I retire, if no one pays attention to me, it is all right.

I hope that before retiring, I will be able to establish a type of mechanism that as much as possible uses rules to govern those actions that now require subgroup leaders to resolve. In the future, if there is some dispute, committees can be used to resolve them, or new rules can be created to handle it. If there should be a subgroup that wants to declare independence, or does not observe the rules, I think there is no harm in it because if there is agreement at the shareholders' meeting, there is nothing left but to accept the situation; the only condition is that the subgroup can no longer use the Acer name.

Any person can be successfully trained

Q: **When selecting a manager, what standards of judgment do you use? Is a total employee evaluation system needed?**
A: Acer is experimenting with such systems, but we are not very familiar with them yet, and we still need more experience with them. Any kind of person can be trained; with honest communication and a generous exchange of each other's experiences, he or she can gradually be given responsibility for a high position.

Q: Can a leader be trained, or are leaders just born with the necessary skills?

A: Morris Chang (CEO of Taiwan Semiconductor Manufacturing Corporation) and I have very different views. He believes that leaders are born, while I believe leaders can be cultivated. In fact, both views are correct. A leader must have certain inborn talents, but these talents may not be as rare as it appears. It may be that because of a lack of the right environment, people with the right talents never get a chance to demonstrate their suitability to be leaders. Therefore, running an organization means providing the best environment possible and training all talented people as potential leaders; this way, there are many people who are possible successors. Designating a single person as a successor carries too high a risk. A leader's biggest task is to train other leaders, and this is the biggest contribution he or she can make to the organization. If all skills and power are transferred to a single person only, there is the danger of the wrong judgment being made in his or her choice of successor.

In the process of leadership succession, the stability of the organization is very important, and time is needed for preparations. Only if a business provides a stable environment can it develop smoothly. When the internal situation is not stable, the board is divided, there is no leadership focus, and a completely different result will arise.

Q: Acer is an incorrigibly changeable organization, constantly developing; even though an organization that is constantly metamorphosing is hard to destroy, it may remain undistinguished. How do you believe its status can be raised?

A: I must admit that Acer is like this, so I have always made efforts to give Acer a more collective identity. Each entity within the Acer Group is metamorphosing, and everyone's mission is to strive to make Acer world-famous, so we all share a single brand. Our goal is "Acerware everywhere"—things done by Acer everywhere. We want to solidify this developing organization as a business of the first rank, achieve an even higher standard, and stand as an equal with other multinational corporations. Acer has distinctive features, such as its long-term perspective and its patience as an organization. Which U.S. company has a CEO who thinks decades into the future? The life span of most U.S. companies is very short, but I believe that an internet organization structure has a chance to turn Acer into a "higher form" of organization in the future.

Internal competition

Q: As subgroups within Acer continue to develop, is there a danger of them fighting over new startups?
A: The Acer Group must institute more stringent standards. I have discovered that the speed at which the group is producing new companies seems a little too fast, and more attention should be put on quality. Once a company has been created it must be nurtured, and through the organization's efforts we allow each new company to be healthy and to develop effectively. For example, when some companies that were less than ideally run were about to go public, our policymaking committee had a meeting and decided to require that all initial public offerings (IPOs) be approved by the committee. We don't evaluate a candidate company only on the basis of its business performance, but also on accounting quality, management integrity, and other basic principles to discern if the company needs to strengthen itself in any areas before going public.

At present, each subgroup has its own territory, and the only areas where there is still some ambiguity that needs to be addressed are businesses related to the Internet and venture capital. Everyone wants to do venture capital investment, but in earning money from venture capital investments, the Acer Group has only one company that has performed with much more distinction than the rest: Acer Venture Capital. We cannot control every action that takes place within the Acer Group, and should there be an "unplanned birth" of a new company that turns out well, we are willing to welcome it as part of the group; natural law works this way, and there is some give and take within the Acer Group.

CHAPTER 12

New People or New Thinking?

RE-ENGINEERING is a normal part of a business's lifecycle, but unsuccessful re-engineering may affect a company's survival. In organizational re-engineering, the company's head is the most important. Without changing the person at the helm, re-engineering is virtually impossible. However, even if the person is not changed, his or her mindset should.

At the end of the 1980s, re-engineering gradually developed into a popular trend in the U.S. Perhaps spurred by talk about Japan becoming number one, U.S. businesses started looking at themselves and came up with re-engineering approaches. Thinking back on those events, it's easy to see that re-engineering is actually just a normal part of a company's lifecycle: if a company goes a long time without evolving and discovers something amiss—targets are missed, mistakes are made in execution, or problems crop up in the organizational structure—it should re-engineer. The longer that a company has been successful, the more effort it takes to re-engineer after the mistakes are discovered. Unfortunately, the success rate for re-engineering efforts is not high, and sometimes there is nothing at all to show for the effort. Many companies fail in their effort to re-engineer and go out of business. This is like the situation when a person has a health problem; if the operation fails, it may affect his or her very survival.

Why re-engineering is needed

Why does a company need to re-engineer? Conditions are constantly changing; external and internal environments are in a state of flux. In addition, from a broader perspective, without challenges or sufficient change, life becomes very boring; the challenge of re-engineering is a very engaging one. What is meant by "conditions" is not limited only to the larger business environment; even the nature of a company's business is a part of what makes up the conditions in which it exists. For example, the Internet has induced a qualitative change in the nature of business in every industry, and re-engineering is needed to react to these changes. To constantly increase their competitiveness, or even their chances of survival, companies need to undertake re-engineering.

In the areas with which I am familiar, I have seen dramatic changes in three industries. One is the PC industry; because of open standards, after the appearance of Microsoft and Intel, the business models of traditional computer companies lost their competitiveness, and the industry became one in which the disintegration approach prevailed. Companies were forced to undertake a series of re-engineering efforts in order to adapt to the resulting changes in the business environment.

The second was the semiconductor industry. The early semiconductor industry was dominated by integrated device manufacturing, with everything being done within individual companies. Thirty years ago, all semiconductor companies just did business by selling a total package, and it was only recently that outsourcing has appeared.

Finally, there is the software industry, which is moving from a traditional model of designing applications and selling packaged software to what is now a development in progress that has yet to reach maturity: the application software providers who rent software, making the use of software akin to the use of electricity or water; that is, you are charged for how much you use. This kind of change will serve as a catalyst for a whole series of further changes in the industry.

The scope of re-engineering

How large is the scope of re-engineering? A change in organizational structure is a visible form of re-engineering, like the division of Acer into global business units from a mixture of strategic and regional business units; however, other than visible changes, there is also a less tangible aspect to re-engineering, and this conceptual re-engineering side is often completely neglected. For example, what are the basic concepts embodied by an organization's system of management? Is it that people are fundamentally good, or inherently only self-serving? Is it about centralized authority or individual autonomy? These things, which are intangible concepts, are at a higher level than organizational restructuring and are too often not given careful discussion. Does everybody agree with the concepts underlying a re-engineering initiative? These are not just slogans to be chanted, but embody vital directions for re-engineering.

A change in business focus is also within the scope of re-engineering. For example, two momentous events in the industry that came out of re-engineering efforts at the companies involved were Intel's elimination of its DRAM business and Texas Instruments' exit first from its computer business and then from its DRAM business. A change in business scope is the most critical aspect of a company's re-engineering. If in putting your efforts into a certain business, you find that the more you do, the more you lose, then it's time to get out of that market. A good example is Acer's exit from the U.S. consumer PC market; this was a very important decision.

A change in business philosophy or an operational workflow is also encompassed by the idea of re-engineering. When Acer undertook its re-engineering, if at a certain point in a workflow there was no value added, then that point in the workflow was eliminated. This led to the organization being transformed into an internet organization, with each "node" being a point at which real value is created.

Resource re-allocation is also very important; when your business has changed, and your organization has been adjusted, the way resources are allocated must also be modified. For example, when Acer wanted to get into the business of digital services, or the components business, key people in the organization had to be deployed in these areas.

STRATEGIES AND PROCEDURES FOR RE-ENGINEERING

In organizational re-engineering efforts, the company head is the most important and must be changed. If in the re-engineering process the company head is not changed, it's virtually impossible to make it work. Many companies' re-engineering efforts only succeed totally after changing leaders. IBM is perhaps the best-known example; in the late

1980s and 1990s, re-engineering efforts under the same CEO, or even the new CEO from within IBM itself, did not achieve the desired results. They were eventually forced to go outside the company and bring in Louis Gerstner before finally succeeding in turning the company around.

There is often no choice but to change CEOs because the original leader will think about how he or she was very successful in the past and was very conscientious, and blame wrong business conditions or a lack of cooperation from employees for the company's problems. If a CEO thinks this way, there is no way to make a fresh start. Many family businesses in Asia are unable to change the company head, and the company gradually declines. This sort of phenomenon is relatively rare in the U.S.: no matter how well a CEO did in the past, he or she is shown the door if that's what the company needs as determined by the company's board.

The re-engineering process is a very long one and requires constant communication and the achievement of consensus. Furthermore, it must be top-down and company-wide. The arrival at a final decision requires the input of the members of the organization, and the common interests of the absolute majority must be considered. Then, after the conceptual direction for the re-engineering has been determined, top-down communication should proceed.

For a re-engineering effort to be successful, things generally have to be simplified, while some tasks can even be thrown out and no longer performed. In Acer's experience, a large organization has to be split up into a dozen or more small organizations, each autonomous and managing itself, and only through this simplification is there a chance to achieve focus.

KEY FACTORS FOR SUCCESS IN RE-ENGINEERING

Of all the key factors in making a re-engineering effort successful, the foremost is that the organization must have a sense of crisis. In general, the larger the organization, the weaker is its crisis mentality. When a large organization re-engineers itself, it must create a crisis mentality, be able to accept change, and win the support of at least the majority of employees.

Even more importantly, top-level executives must make commitments, and their actions must be aggressive, utilizing avenues for communication, building consensus, and creating shared interests to give the members of the organization confidence before any concrete actions are taken. The situation can be likened to a patient before surgery needing to be told the reason for the illness and what will be done to treat it, giving the patient confidence in the surgical procedure.

The process of re-engineering is extremely lengthy, so there must be a clear blueprint, as well as some measures that can easily achieve visible results in a short time—just as a post-operative patient who needs six months to fully recover still needs to see some signs of recovery in the first few weeks, as otherwise he or she will lose confidence. Likewise, with some small successes achieved very quickly as encouragement, the members of an organization can develop confidence in the process of re-engineering.

Proper understanding of what re-engineering is

Before undertaking re-engineering, an organization must first understand just what re-engineering is and what it entails. Acer re-engineered when it was a billion-dollar company, and the whole process took about three years. First, the preparations took several years, and after concrete actions were taken, it took one or two years for visible results to be apparent. During this time, the correct direction must still be sought and confirmed, and dissenting viewpoints handled—this requires considerable time. IBM spent more than five years; I've heard that Philips has been talking about moving from SBUs and RBUs to GBUs for ten years but hasn't completely made the transition. Of course, most of the work has been completed, but resistance from many ingrained attitudes and ideas needs time to break down, and this accounts for the low success rate of re-engineering efforts in general.

Re-engineering has another distinctive characteristic, which is that there will be pain initially, but after this pain things will gradually become smoother. Re-engineering cannot be limited to just talk, as is the case with most Japanese companies; if no action is taken, obviously there is no chance of success. In the early 1990s, the Japanese saw that U.S. businesses were re-engineering and repeated some slogans about re-engineering. Each time I go to Japan, I always hear that Japanese companies are eager to re-engineer themselves, but the results have been disappointing.

IBM's re-engineering

IBM was the embodiment of the vertical integration approach in the computer industry. In its earlier years, IBM was the largest semiconductor company and the largest software company; it made all its own products and didn't rely on any other organization. I once spoke with a vice-chairman at Texas Instruments, who told me that in the early 1990s he had tried everything to try and persuade IBM to alter its traditional semiconductor business and instead work with Texas Instruments, but IBM categorically refused. After IBM's new CEO, Louis Gerstner, took the helm, IBM suddenly changed and began looking to outsource, and this even extended to its computers. Besides this, IBM's technology was opened up for use by other companies, as a result of which our semiconductor company, TI-Acer, was able to begin some cooperative ventures with IBM. All of this was made possible by an important change in thinking from the CEO.

In 1997, I went to IBM to speak with Gerstner, and the first thing he said on seeing me was, "IBM's R&D division has a lot of things going on, but I don't know what they're doing. Take a look for yourself, and if you see anything you want, take it." We of course had to pay some money to license IBM technology, but because of what Gerstner said, I actively pressed for an evaluation of the possibility of using IBM technology in areas that Acer was getting involved in, such as LCDs.

In the end, IBM opened its technology to the entire industry. This approach is correct; however, making such an enormous change was not easy. Two years ago, IBM reorganized, and Gerstner announced that in the future the parts of IBM with the highest growth would be those selling technology, and income from technology is far higher than previously.

Aside from technology, IBM provides services to other companies and is willing to do this even for competitors such as Dell. IBM has already established a good infrastructure; if it is only used by IBM alone it's not an efficient use of resources, and the full returns possible from technology are not realized, so it should be freed for everyone's use. IBM's main business in the future will likely not be hardware, but the services and e-commerce that it is aggressively pushing now.

RE-ENGINEERING OF JAPANESE COMPANIES

As mentioned, Japanese companies began talking about re-engineering in the early 1990s, but few have taken substantive action. This situation is related to Japan's social structure and culture, and is not purely a lack of will. The Japanese are most proud of their lifetime employment system; in undertaking re-engineering, this kind of culture and system becomes an enormous obstacle.

Many Japanese companies constantly emphasize improvement, and no one thinks more actively about how to improve than Japanese companies. Japan has produced many writers in this area, and many Taiwanese ideas about quality control originated in Japan. This way of thinking is one of the key reasons for the success of Japanese business.

The problem is that business conditions are changing too quickly so that even if you do something well, or execute it perfectly according to yesterday's rules, your results today may be sub-standard. Meanwhile, other companies that are better tuned to current market conditions may execute it less well, but still achieve better results. If in terms of current conditions you are doing the right thing, you don't necessarily have to execute it perfectly, but if you are doing the wrong thing, perfect performance is useless—and this in a nutshell is what is keeping Japanese business as a whole from performing up to its potential. Add to this the fact that Japanese business is inferior in terms of speed and flexibility, and they're not open-minded enough to accept change. In 1987, I went to Japan to talk about DRAM, and I went to all the Japanese companies in this industry, but not a single one agreed to transfer technology. For the most part, the Japanese are not able to think like Texas Instruments and share technology.

In Japan, Sony can be considered an exception. Sony, in fact, in some sense doesn't count as a Japanese company because founder Akio Morita spent his early career in New York and made the U.S. market Sony's focus. Sony's center and main capabilities are situated in the U.S., and its position in Japan is definitely not as high as Panasonic, even though its reputation globally is much higher. In 1996 or 1997, reform within Sony was big news in Japan, even though the main action of Sony's board was to hire someone from the outside to serve as chairman, and at the same time reduce the size of the board from more than twenty people to less than twenty. The board of a company is only concerned with strategy and not with company operations, so Sony's action did not even qualify as a full-blown re-engineering initiative; and yet Sony's approach created a furor in Japan because even its limited action was a break from Japanese tradition.

After Sony, virtually every company started paying more attention to Web business opportunities and made corresponding adjustments; however, the only company to actually successfully transform itself was Sony. The Japanese companies today, such as Canon, all set up Web business divisions, but there is virtually no Japanese business that has re-tooled itself company-wide in response to the Internet revolution.

Acer's re-engineering

Acer's first large-scale re-engineering took place in 1992. Before that, in 1989, we had initiated efforts to flatten the organizational structure and create a system of "rest stations"—positions where older employees who had reached a bottleneck in their careers could "rest" before making a fresh start, while younger employees could take over tasks where speed and energy were a premium. Later, we very quickly instigated an early retirement plan.

Still, at the beginning of the 1990s, the company's performance started to decline, and we felt a great deal of pressure. We came up with the concept of a global brand with a local touch; all group companies would use the same brand, but each company itself would be autonomous. I constantly stressed both globalization and localization, in both their geographical and conceptual senses, and this idea is still effective even today. As for organizational structure, at the time we were using a client–server approach; it was combined with the idea of a global brand with a local touch to become an internet organization. In terms of workflow, we utilized the fast-food manufacturing and sales model.

After settling on a re-engineering framework, we commenced many communication initiatives within the company to explain the principles behind it. I reiterated ideas such as the smiling curve mentioned earlier. The "21 in 21" initiative and the NT$200 billion revenue target for the year 2000 were also presented at that time.

Beginning with this re-engineering program and continuing to the present, each year Acer holds a large forum to discuss the encouragement of internal startups. My hope is that the current Acer Group and subgroups can all be powerful in their domains without any single one trying to dominate the others.

Table 12.1 Re-engineering of Acer (3)

High growth in revenue and profit

	Revenue (US$m)	Growth rate (%)	Profit (US$m)	Growth rate (%)
1993	1,902	51	38.6	2,436
1994	3,220	69	118.3	207
1995	5,825	81	202.9	72

© 2000 Aspire Academy. All rights reserved.

Looking at the process of re-engineering, it can be seen that if it's done correctly, there can be 50%, 70%, or even 80% growth (see Table 12.1 above). In 1986, I set a revenue target for 1991 of US$1 billion. At the time, I planned on growth slowing from an annual 25% to 20% or 15% because I felt the larger the company was, the more difficult growth would be. This conclusion seemed to be reasonable, but in fact it was wrong. At the time, I saw Compaq's incredible growth, and recently saw Dell's growth, and realized the notion that the bigger a company is the slower its growth is not necessarily correct. The key is whether you are doing the right things, and if you are, growth is not limited.

In the early stages, we doubled in size every year; compared to our competitors we did more of the right things, and the result was our rapid rate of growth. I have constantly thought about the situation when objective rates of growth have slowed; if you want growth rates relatively higher, you must achieve the same things with fewer resources. If a company is only able to achieve single-digit growth, it's best if the number of people needed is reduced by 10%. In this way, double-digit growth on a per-employee basis can be achieved. Without this high rate of growth, a company just is not competitive.

ACER'S RE-RE-ENGINEERING

A company should undertake re-engineering once every five years. As a kind of follow-up to its earlier re-engineering efforts, in December 1997 Acer began a process of re-re-engineering. In 1992, the company's condition was poorer, and in 1997 it was better, but I still felt that the company was not in a position to make the most of its opportunities. During this re-engineering effort, I gathered together key managers from around the world and held a two-day meeting at which the following conclusion was reached: if the company did not strengthen itself in certain areas, it could no longer stay competitive.

For mass market commodities, such as the PC has become, to be competitive, an end-to-end globalized logistical workflow had to be considered, otherwise risks and costs would grow, and competitiveness would suffer greatly. To attain this objective requires an effective information system infrastructure. Additionally, to fight this battle, if no thought were given to brand management, something would be lacking, so we instituted the total brand management initiative. The biggest source of differentiation between companies, and the key to survival in the PC industry, is not products but services for customers. In 1997, in order to enhance service and strengthen the brand, we established a brand business unit to concentrate on building up the brand.

At the 1997 re-engineering meeting, we discovered that if SBUs and RBUs were not integrated as GBUs, it would not be possible to effectively implement an end-to-end operational workflow. Therefore, our Singapore and Mexico companies, which had already gone public, had to prepare to delist. Because our U.S. and European companies had not gone public, it was easier to integrate them as GBUs.

We also slowly transformed our client–server organization structure into an internet organization structure, the concept of which was presented in 1999. Acer established a new intellectual property and digital services subgroup. In the past two years, various departments within the company have held meetings to discuss changes in the corporate culture, modifying it into one that is completely customer-centric. Acer is so large as an organization that changing employees' mode of thinking to make it customer-centric is definitely a major undertaking.

Additionally, the company's positioning and its direction have changed from PC clone-maker to Web lifestyle-enabler; the entire group and each subgroup serves as an enabler for Web applications. Today, we are seeing some initial results from the re-engineering, and they have been encouraging thus far.

REACTIVE AND PROACTIVE RE-ENGINEERING

A company should actively pursue re-engineering and not merely accept a reactive form of re-engineering. Re-engineering is a normal part of a business's lifecycle, and if it is delayed too long, it not only adds to the difficulty of re-engineering, but at the same time causes an inability to effectively utilize resources, with the result that performance will not meet expectations. Therefore, it is critical for a company to initiate re-engineering before decline sets in.

If the re-engineering direction is chosen correctly, a small organization can generally see results in three to six months, a medium-size organization needs a year, while a large organization may require two or three years before results become apparent. If the direction is wrong, the problem may not be seen for three years in a large company, while a small company will show indications of the error in three months to half a year.

Therefore, a large organization must possess a bold way of thinking; if there is any sign of a downturn or if company performance is less than ideal, attention must be paid and adjustments made immediately. To have this kind of sensitivity sometimes requires a reliance on experience. Intel's Andrew Grove once said that a company may not always be able to take advantage of turning points, but must regularly initiate re-engineering. For the sake of a company's competitiveness, re-engineering should be an aggressive maneuver and not a reactive, defensive move.

Conclusions

Initiating and leading re-engineering efforts is one of management's most important functions, and a company must regularly re-engineer itself. Re-engineering is not only an organizational change; the larger an organization, the more drastic the actions that must be taken and the longer the time required. During the process of re-engineering, a new CEO is most important, regardless of whether this means actually changing CEOs or just changing the mindset of the current CEO. An industry requires re-engineering; companies and departments need to re-engineer; every person should also re-engineer. Begin the re-engineering process with smaller groups within the organization, and when the larger environment changes, the whole organization will be adapted. Re-engineering requires establishing your team's confidence: without confidence in the overall process, there will be many actions that people will be afraid to take, and this will cause considerable delays.

In 1994, *Business Week* saw me as "turning defeat into victory." I felt a little frustrated by this description—I hadn't failed, so how could I "turn defeat into victory"? People within Acer were very confident: they were clearest about Acer's condition. However, a company does not only live in the hearts of its employees, but it must also pay attention to the level of confidence the outside world has in it. Customers, banks, suppliers, employees' families, shareholders—they are all people outside the company whose level of confidence in the company's re-engineering is important.

Discussion

Shorten the period of fear

Q: Many companies wait until obvious problems arise before they want to re-engineer. Re-engineering requires confidence and trust. How can one proceed in order to minimize the period of fear and uncertainty? During re-engineering, how is the allocation of power handled?

A: An organization should have a sense of crisis at all times, but naturally the shorter the period of acute fear the better. To quickly dissipating any feelings of unease, the company must clearly communicate its blueprint for the future, explain what actions are to be taken, and announce whether there will be further initiatives. Otherwise the company's employees will be caught in an atmosphere of fear, and their performance will suffer as a result.

The first part is the preliminary framework, including protection of the company itself, which is in the interest of everyone at the company, while the second part is personal interests. However, the interests of individuals must take a back seat to the good of the company. If the company's interests cannot be accommodated, it's bad for both the company and oneself, but the company must also consider the future of individuals. This might mean, for example, adequate early retirement plans, and for such eventualities the company must have sufficient resources.

Find options for talented people

Let me explain using the actual case of Acer America's re-engineering initiative. Dr Ronald Chuang at Acer America was a most outstanding talent, but couldn't produce the needed results in the computer industry—business conditions in the PC industry are very tough. I replaced him as CEO, and at the same time provided US$40 million for him to invest in the U.S. Although he could accept this arrangement, he was still very upset, thinking himself a failure. However, the money he is earning for Acer now is higher than in the past. Providing for his future was a very important step; whether this was giving him some money to enjoy retirement with or letting him do another job, we had to arrange something for him. If a company is very healthy, there will be many options; and in the information technology industry, there are so many alternatives—of which starting a new business, as we did, is one.

Hong-yi Lu at Acer's Singapore operation now handles venture capital business after the delisting. For him, this is a new challenge. Therefore, when making preliminary arrangements for a group with shared interests, and making personnel adjustments, one must find other options for the people whose interests are in conflict with those of the larger group. Even if the option chosen turns out to be staff cuts, resources are needed. In Acer's early retirement initiative in 1990, its severance package was better than that required by labor laws. After the employees left the company, we wrote letters of introduction for them; and within a few months they found new jobs. So one must assist people in transition to find a new road, and this way the unease in the organization can be reduced and performance will also be raised.

Many companies use the name re-engineering as a cover for staff cuts. For example, it might use a company's move to a new location to encourage employees to leave, and avoid the need to pay severance fees. This kind of re-engineering will arouse suspicions and will not win the support of most employees. Each time Acer moves, it pays out several years of special compensatory payments and does not use the move as an excuse to cut staff. Nearly all the people we have had to let go would not have anything negative to say about Acer—we took care of their interests.

Establish thought modes that allow radical change

Q: **Can a CEO completely change his or her way of thinking, as you say is necessary for a company to successfully re-engineer itself?**

A: In Taiwan and many other Asian countries, it's very difficult for a company to change its CEO, especially in family-run companies because of power and "face" issues. Acer is quite exceptional in this respect, with its ability to change CEOs whenever necessary, even though this is still a very difficult process to carry out.

We first tell the group CEOs that they are professional managers; if they don't perform well, they will be replaced. Even I have submitted a resignation. The only caveat is that if the situation has not deteriorated to the point where there is no choice, we won't casually change CEOs. When Acer switches CEOs, we make sure the CEO doesn't lose face.

I can only repeat that if a CEO cannot constantly adjust, and if he or she cannot be replaced, this company will inevitably decline. There are isolated cases of companies that didn't change CEOs and made a comeback after several years of decline; the reason is that after several years of poor performance, the CEO finally came to the realization that some radical changes had to be made or there would be no opportunity of proving himself or herself again.

When a CEO seeks to transform thinking patterns, he or she must find an approach for creating an environment conducive to change. Taking the Acer Group as an example, on the very day it was founded, I established an environment that encouraged radical changes in ways of thinking. I said to everyone, if I could not lead Acer's development, then someone who could would have to be found. This was not only directed at myself, but also at other managers. I might need to cut myself out of the picture, and other CEOs might also have the same done to them. From the first day of Acer's existence, I created this sort of environment. If I needed to cut myself out, I didn't have to provide any other options for myself and would just resign. However, if other people need to be cut out, some other options for them must be found.

The dilemma of centralized versus distributed power

Q: **Acer is a multinational corporation; how can a balance be struck between centralized and distributed authority?**

A: Some initial conclusions have to be reached on the issue of centralized versus distributed authority, and then as times change, these can be adjusted. Basically, we first formulate an approach, then use problems that happen to arise to reflect on the approach, and sometimes even deliberately let things slide a little just to see what happens. For example, Acer's basic principle is to allow local autonomy, and before I take back authority from overseas operations, I allow them to develop problems: unless problems have actually occurred, overseas leaders won't be willing to relinquish power. Only by making them feel that giving power back to headquarters is to their own advantage will they support headquarters having control.

Fortunately, an approach based on distributed authority is suited to the industry Acer is in. For example, the scope of R&D is very broad, and greater autonomy for individual operations is more effective, so we distribute authority to local operations. However, for manufacturing, centralized authority is more effective: manufacturing benefits from economies of scale, and its possibilities for adding value are fewer.

In the past few years, Acer has been discussing internally whether or not to establish a central R&D center, as IBM and Bell Laboratories have. However, I have never been able to accept the view that centralized authority is more effective than distributed authority in the area of R&D. Brand image strategies and positioning should be made by a centralized authority, while the responsibility for brand applications should be distributed as much as possible. Also, marketing should be done by autonomous units.

From the point of view of sales, each company in the Acer Group is autonomous. As for global strategies for the same product, in the past this was done by autonomous local operations, but now the decision-making is centralized. The reason for this change is that for end-to-end operational workflows, from R&D all the way to customer service, management by a single person is more effective. Besides this, intangible items, such as training, should be handled as much as possible by a centralized authority; the more localized tasks should be left as much as possible to the local operations to take charge of.

Communication and consensus are most important

Q: From a company's establishment to its first re-engineering, approximately how much time is needed?

A: In our industry, there are around five years in each cycle, at the end of which re-engineering is needed. However, it is very easy to re-engineer when an organization is small, and the re-engineering process may not even be perceptible. As a result, this may not leave any scars. In Acer's case, more than ten years had passed from the company's establishment to its first "visible" re-engineering; when the organization was still small in scale, we re-engineered every year without anyone noticing.

Now, many Web companies are likely to completely change their business plans six months after they have established themselves. Faced with changes in the business environment, a company must change as well, to develop a different way of thinking. If a company is using up many resources and much effort, and the quantity of talent within the company is not less than at competitors, and yet the company is still performing poorly, the time for re-engineering is at hand.

Q: **You emphasize that Acer will use CEOs under forty years of age. How will these young CEOs lead older employees in re-engineering initiatives?**

A: This is easier in the U.S., but in Asian countries it's virtually infeasible. The internet organization structure that Acer has designed for itself needs to solve this kind of problem. If a CEO is forty years old, his or her operational team will in principle all be under forty years of age. Older employees can work in training or do venture capital investments or other areas of business where broad experience is especially valuable.

The most important things in re-engineering are communication and consensus, and barriers to communication will create obstacles to the re-engineering effort. I also want to especially emphasize that Acer is very lucky to be part of an era where business opportunities are so plentiful; with the digital economy, these theories are perhaps workable. In the Internet age, the speed at which the old is overtaken by the new is rapid; how to open one's mind and communicate before an action is taken, and not rupture the social culture governing the relations between elders and the younger generation while still taking advantage of new opportunities—these are all issues that demand thought, particularly in Asian companies.

Being a CEO in different spheres

Q: **The key to re-engineering is the CEO, and yet it is difficult for a CEO to listen to the recommendations of others, therefore subordinates cannot exert much influence. Under such circumstances, what course of action should be taken?**

A: There are many levels in society, with a CEO for each level. You might be the CEO of your family or the CEO of your own life. While you are lamenting that you cannot exert any influence on your company's CEO, any person can make the same kind of lament about not being able to affect what the government does. We can only use our resources effectively and exercise whatever power we have to make the things within our sphere of influence run as smoothly as

New People or New Thinking?

possible. When there is no room for you to maneuver in a company, you are definitely entitled to change companies or start your own.

A company's success is multi-faceted. If a company is earning money, satisfying employees, satisfying customers, or creating a favorable image, there is no reason to re-engineer. However, in today's business environment, using the same business model for five years or ten years and still remaining successful is nearly impossible. There is no line of business in the entire world in which you can do the same thing year after year and survive.

Japan provides an example. In 1986, I went to Japan to try and persuade companies there to work with Taiwan on an equal basis, but at the time they flatly refused. But now, in the high-tech industry, Japan has some sectors that might do well to study what Taiwan has done. When Japan was successful, most of its companies would not listen to what I had to say; but after several years of hard times, they find it easier to accept foreigners' viewpoints. Who could have imagined that the notion of Japan as number one would disappear so quickly? This is a consequence of an unwillingness to re-engineer once they had become successful, and Japanese companies must accept responsibility for what has befallen them.

Although I can't influence many things, this inability only applies to the short term. Over the long term, I can have an impact. The managerial system used in Taiwan's information technology industry today is worlds away from that used in traditional industries. Because people in the high-tech industry do things with greater impact, they are developing confidence, even though they are still in the minority. As they continue to develop, they are exerting an influence on traditional industries as well. I am willing to try and exert an influence through my own efforts, though the impact might not be felt for ten or twenty years.

A sense of external changes benefits re-engineering

Q: **How do you present Acer's re-engineering initiatives? If during the process some error is discovered, how do you make adjustments?**
A: The role of a CEO is to lead re-engineering. Re-engineering is necessary because changes in the external environment are too marked to allow a company to continue with the status quo; however, if attention is not paid to them, these changes may not be noticed, and there will be no way to initiate the required re-engineering.

The reason why I am able to sense the need to initiate re-engineering at Acer is that I do a lot of things that are not directly related to business; if all I did was directly related to business, paying attention only to issues at the company, I would have no sense of the changes going on outside. Basically, my understanding of the world is gained through many meetings and discussion workshops, where I can hear what people are concerned with. In the same fashion, a CEO must be sensitive to changes within the company, and I can sense whether people have confidence or not, whether morale is high or low, whether there are many complaints or few. I must have a grasp of these factors; even if I don't know all the details, I must be able to perceive the changes as this is the basic preparation that a CEO must undergo before initiating a re-engineering effort.

The re-engineering process is adjusted as circumstances change. With the current re-re-engineering, we originally wanted to establish at group headquarters a line-of-business manager, cutting across all Acer subsidiaries, but after only three months we knew it had failed. The reason was that this approach contradicted our basic principle of each company being independent and autonomous, and headquarters being only virtual. Re-engineering initiatives cannot be overbearing; until a general consensus is achieved, it cannot be instituted in an autocratic manner. Undertake the re-engineering after consensus is reached, but once begun, insist on its continuation even in the face of difficulties.

In re-engineering, you must depend on yourself

Q: Some companies re-engineer themselves on their own, while others use the services of a consulting company. What is Acer's experience in this area?

A: The first time that Acer re-engineered it sought out McKinsey & Consultants, but after talking for a while we stopped and did the re-engineering ourselves. The second time we re-engineered, which primarily involved a re-working of our operational workflow, we used an outside agent (again McKinsey & Consultants), with the main consideration being a very thorough, detailed re-engineering approach. We needed professional expertise in order to establish a structure and create a good system.

To change a corporate culture requires all the managers within the company to perform a top-down adjustment, and you must rely on yourself, and not an outside agent, for this. I led the initiatives related to internal management in our second re-engineering, such as becoming more customer-centric, emphasizing intellectual property, putting more focus on services, as well as the creation of subgroups. Acer is different from other companies: it has no trouble acknowledging that it needs to change and accepts this process.

Hiring an outside consultant is done simply to verify that there is no problem with the general direction we have chosen, while we do the actual execution of the re-engineering ourselves. When employees are willing to change, an outside consultant is not strictly necessary. However, when people within the company resist change, then even if the head of the company wants to change, he or she will find obstacles to the successful execution of the re-engineering. At such times, the CEO can take advantage of an outside consultant to help with communication and persuasion. Still, an outside consultant is only a source of assistance, and cannot be the primary driving force—in re-engineering, you must still depend on yourself.

References

Thomas J. Peters and Robert H. Waterman, Jr. *In Search of Excellence: Lessons from America's Best-Run Companies*, New York: Warner Books. 1988.

Stan Shih with Wennie Lin. *Me Too Is Not My Style,* Taiwan: Commonwealth Publishing, 1996.

Index

A

Acer 12, 14, 19, 21, 28, 31, 34, 376, 43, 44, 46–48, 50, 52, 53, 55, 60, 73, 75, 76, 79, 82, 83, 87, 9–98, 102–106, 109, 114, 117–119, 120–124, 126, 127, 130, 131, 132, 136–137, 138–139, 141–145, 147, 148, 149, 150, 151, 153, 157, 159, 160–162, 165, 169, 170–173, 179, 180–183, 185, 189, 191, 197, 198, 199–202, 204, 206–207, 208–209, 215, 218–220, 222–223, 224, 225, 226, 228
   Acer America 189, 193, 222
   Aspire Park 138
   Micro Professor 106, 121, 157, 172
Asia viii, xii, 8, 12, 16, 23, 24, 57, 59, 60, 61, 62, 64, 70, 71, 74, 75, 80, 81, 84, 85, 90, 91, 92, 94, 97, 98, 101, 109, 113, 114, 116, 120, 121, 126, 127, 128, 135, 145, 149, 150, 151, 164, 173, 183, 191, 204, 214
Asia Management Academy 136
Asian financial crisis 139

B

B2B e-commerce 30, 63, 83

Barnes & Noble 35
brand management ix, 16, 17, 51, 123, 130, 153, 161, 162, 172, 220
business philosophy 64, 96, 180, 186, 188, 191, 193, 203, 204, 213

C

capital markets xii, 89, 92–93, 101
chaebol 48
China
   Shanghai 137, 138, 146, 184
   Shenzhen 137
   Suzhou 137
Compaq 85, 122, 159, 166, 167
contrarian thinking 99, 107, 109–110
corporate culture ix, 10, 19, 43, 50, 51–52, 55, 89, 94–95, 96, 97, 177, 184–189, 190–193, 202, 220
customer-centric viii, 3, 8–11, 12, 77, 123, 220

D

Dell 6, 72, 75, 122, 159, 160, 216
disintegration and super-disintegration vii, viii, 3, 6–7, 9, 17, 22, 26, 27, 29, 33, 37, 38, 39, 70, 98, 114, 115, 119, 133, 134, 200, 212

distributed management 55, 98, 181, 182, 184, 199
Dubai 139
dynamic random access memory (DRAM) 8

E

e-commerce 22, 24, 25, 27, 30, 32, 33–34, 63, 83, 216
  B2B 30, 63, 83
  B2C 27, 30, 32, 33, 34, 63, 83
Europe xi, xii, xiii, 78, 80, 85, 115–116, 119, 120, 121, 128, 131, 135, 138, 143, 162, 165, 166, 167, 191

G

global depository receipt (GDR) 140, 141
Germany 131, 138, 156
global business units (GBU) 53, 123–124, 130, 212
globalization xii, xiii, 17, 79, 80, 113–119, 120–122, 126, 133, 134, 140, 163, 218

H

Hewlett-Packard (HP) 11, 123, 124
Hong Kong 23, 28, 80, 137, 157, 166
Hsinchu Science-Based Industrial Park (HSIP) 91, 121, 177, 182, 200
human-touch bits 21, 60, 179

I

IBM
  Louis Gerstner 122, 214
initial public offering (IPO) 209
internet Organization Protocol (iOP) viii, 51, 57
innovalue viii, 69, 76, 77, 100, 158
innovation
  environment viii, 65, 80–81, 89, 93–94, 97, 100, 105, 106
    importance of market scale 89, 97, 138
    influence of capital markets 92
    influence of corporate culture 94

influence of social culture 93
Intel
  Andrew Grove 221
intellectual property 5, 11, 12, 32, 60, 89, 93, 97, 105–106, 123, 147, 220
Internet viii, 3, 5, 6, 7, 8, 13, 14, 16–17, 18, 21, 22, 23–24, 25, 26, 27, 28, 29, 31, 32, 34, 35, 38, 45, 47, 48, 50, 51, 55, 57, 59, 60, 75, 76, 84, 95, 96, 101, 164, 178, 179, 184, 209, 212, 218, 226
internet organization xii, 37, 38, 41, 43–57, 96, 98, 102, 117, 119, 123, 151, 182, 199, 202, 205, 209, 213, 218, 220, 226
Italy 120, 131, 132, 164

J

Japan xii, xiii, 18, 70, 74, 91, 93, 104, 116, 117, 119, 121, 133, 144, 151, 168, 170, 183, 211, 215, 217, 227

K

knowledge economy xi, 38, 39, 40, 43, 48–49, 50, 53, 61, 63, 64
Korea 18, 93, 140, 144, 151, 170, 183

L

leaders
  successorship 204
Legend (China's PC vendor) 159

M

Malaysia 136, 141–143, 145, 148, 149, 150
manufacturing ix, xi, 6, 12, 25, 28, 60, 72, 77, 79, 80, 84, 85, 114, 120, 121, 123, 126, 127, 128–129, 133–139, 141–147, 149, 154, 158, 164, 167, 171, 186, 188, 196, 200, 201
marketing viii, xii, 11, 12, 22, 28, 32–33, 34, 62, 71, 72, 74–75, 80, 84, 85, 114, 117, 120, 121, 123, 125, 126, 128, 129, 138, 139, 143, 145–146, 149, 150, 158, 160, 165, 168–169, 172, 200, 201, 202
Mexico 127, 137, 142, 148
Microtek 121

Index

Middle East 13, 139

N

national operations (NO) 125
Netherlands, the 138
New Economy, the 13, 16, 17, 19, 21, 23, 61

O

OBM (own-brand manufacturing) 145, 153, 154, 155, 156, 167–168
ODM (original design manufacturing) 9, 28, 115, 131, 144, 145
OEM (original equipment manufacturing) 9, 16, 28, 115
organizational structures
    client–server organization 98, 201, 220
    decentralized organization 5–6
    hierarchical organization 196, 199
    network organization 196, 197
organization protocols (OP) 51, 57

P

peripheral products
Philippines, the
    Subic Bay 136, 142
product-centric 3, 8

Q

quality and quality control 162, 171, 172, 217

R

R&D 25, 80, 82, 84, 101, 120, 121, 123, 126, 128–129, 137–138, 144–145, 150, 186, 201, 216, 225
re-engineering vii, ix, 122, 198, 199, 201, 202, 205, 211, 212, 213, 214, 215, 216, 217, 218, 219, 220, 221, 222, 223, 225, 226, 228
regional business units (RBU) 123, 212
regional operation (RO) 123, 124, 125

S

service
    customer service 166

Silicon Valley 90, 91, 97, 137, 138
Singapore 125, 139, 142, 148, 220, 223
six-sigmas concept 172
smiling curve viii, 59, 218
software 24, 25, 32–34, 59, 60–65, 82, 90, 212, 216
Sony
    Akio Morita 217
startup 97–98
strategic business unit 52, 123

T

Taiwan vii, xi, xii, 4, 17, 23, 25, 28, 33, 48, 59, 61, 64, 70–72, 76, 80, 81, 84, 85, 92, 93, 97, 101, 105, 116, 119, 120, 121, 125, 128, 135–139, 141–143, 146, 148, 149, 150, 155, 156, 157, 161, 162, 165, 167–169, 170, 177, 183, 187, 189, 197, 200, 208, 223, 227
Taiwan Semiconductor Manufacturing Corporation 208
Texas Instruments 132, 216, 217
total brand management (TBM) ix, 123, 161, 162, 172, 220
total quality management (TQM) 123
training 6, 62, 65, 86, 96, 114, 120, 126–128, 132, 135, 146, 167, 170, 184, 191, 196, 198, 199, 203, 204, 208, 225, 226

U

United States xi, xii, xiii, 5, 6, 7, 11, 13, 18, 23, 28, 30, 31, 33, 55, 56, 59, 60–63, 71, 73, 75, 76, 78, 80–83, 85, 90, 91, 97, 99, 100, 101, 103, 105, 106, 108, 114–119, 120, 121, 124, 125, 126, 128, 129, 130, 131, 133, 136, 137, 138, 142–146, 149, 150, 151, 153, 156, 157, 162, 164–169, 170–173, 178, 180, 183, 186, 188, 189, 191, 199, 204, 209, 211, 213–215, 217, 220, 223, 226

V

value
    creation of 49, 134
    value chain viii, 26
venture capital 56, 91, 92, 101, 138, 147, 202, 209, 223, 226

virtual dream team 6, 45, 83–84, 195
virtual integration 7
vision viii, ix, 15, 31–32, 35, 59, 60, 62, 96, 132, 172, 177–184, 188, 190, 191, 193, 198, 203

W

Web business 22, 23, 30, 31, 35, 218